Human-Robot Interaction Control Using Reinforcement Learning

Human-Robot Interaction Control Using Reinforcement Learning

Wen Yu
CINVESTAV-IPN

Adolfo Perrusquía
CINVESTAV-IPN

IEEE Press Series on Systems Science and Engineering
MengChu Zhou, Series Editor

WILEY

Published by John Wiley & Sons, Inc., Hoboken, New Jersey.
Published simultaneously in Canada.

For general information on our other products and services or for technical support, please contact our Customer Care Department within the United States at (800) 762-2974, outside the United States at (317) 572-3993 or fax (317) 572-4002.
Wiley also publishes its books in a variety of electronic formats. Some content that appears in print may not be available in electronic formats. For more information about Wiley products, visit our web site at www.wiley.com.

Library of Congress Cataloging-in-Publication Data is applied for
Hardback: 9781119782742

Cover Design: Wiley
Cover Image: © Westend61/Getty Images

Set in 9.5/12.5pt STIXTwoText by Straive, Chennai, India

10 9 8 7 6 5 4 3 2 1

To the Wen Yu's daughters: Huijia and Lisa
To the Adolfo Perrusquia's parents: Adolfo and Graciela

Contents

Author Biographies *xi*
List of Figures *xiii*
List of Tables *xvii*
Preface *xix*

Part I   Human-robot Interaction Control   *1*

1   **Introduction**   *3*
1.1   Human-Robot Interaction Control   *3*
1.2   Reinforcement Learning for Control   *6*
1.3   Structure of the Book   *7*
         References   *10*

2   **Environment Model of Human-Robot Interaction**   *17*
2.1   Impedance and Admittance   *17*
2.2   Impedance Model for Human-Robot Interaction   *21*
2.3   Identification of Human-Robot Interaction Model   *24*
2.4   Conclusions   *30*
         References   *30*

3   **Model Based Human-Robot Interaction Control**   *33*
3.1   Task Space Impedance/Admittance Control   *33*
3.2   Joint Space Impedance Control   *36*
3.3   Accuracy and Robustness   *37*
3.4   Simulations   *39*
3.5   Conclusions   *42*
         References   *44*

4   **Model Free Human-Robot Interaction Control**   *45*

4.1   Task-Space Control Using Joint-Space Dynamics   *45*

4.2   Task-Space Control Using Task-Space Dynamics   *52*

4.3   Joint Space Control   *53*

4.4   Simulations   *54*

4.5   Experiments   *55*

4.6   Conclusions   *68*

      References   *71*

5   **Human-in-the-loop Control Using Euler Angles**   *73*

5.1   Introduction   *73*

5.2   Joint-Space Control   *74*

5.3   Task-Space Control   *79*

5.4   Experiments   *83*

5.5   Conclusions   *92*

      References   *94*

Part II   **Reinforcement Learning for Robot Interaction Control**   *97*

6   **Reinforcement Learning for Robot Position/Force Control**   *99*

6.1   Introduction   *99*

6.2   Position/Force Control Using an Impedance Model   *100*

6.3   Reinforcement Learning Based Position/Force Control   *103*

6.4   Simulations and Experiments   *110*

6.5   Conclusions   *117*

      References   *117*

7   **Continuous-Time Reinforcement Learning for Force Control**   *119*

7.1   Introduction   *119*

7.2   K-means Clustering for Reinforcement Learning   *120*

7.3   Position/Force Control Using Reinforcement Learning   *124*

7.4   Experiments   *130*

7.5   Conclusions   *136*

      References   *136*

8   **Robot Control in Worst-Case Uncertainty Using Reinforcement Learning** *139*

8.1   Introduction   *139*
8.2   Robust Control Using Discrete-Time Reinforcement Learning   *141*
8.3   Double Q-Learning with k-Nearest Neighbors   *144*
8.4   Robust Control Using Continuous-Time Reinforcement Learning   *150*
8.5   Simulations and Experiments: Discrete-Time Case   *154*
8.6   Simulations and Experiments: Continuous-Time Case   *161*
8.7   Conclusions   *170*
         References   *170*

9   **Redundant Robots Control Using Multi-Agent Reinforcement Learning** *173*

9.1   Introduction   *173*
9.2   Redundant Robot Control   *175*
9.3   Multi-Agent Reinforcement Learning for Redundant Robot Control   *179*
9.4   Simulations and experiments   *183*
9.5   Conclusions   *187*
         References   *189*

10   **Robot $\mathcal{H}_2$ Neural Control Using Reinforcement Learning** *193*

10.1   Introduction   *193*
10.2 $\mathcal{H}_2$ Neural Control Using Discrete-Time Reinforcement Learning   *194*
10.3 $\mathcal{H}_2$ Neural Control in Continuous Time   *207*
10.4   Examples   *219*
10.5   Conclusion   *229*
         References   *229*

11   **Conclusions**   *233*

A   **Robot Kinematics and Dynamics**   *235*
A.1   Kinematics   *235*
A.2   Dynamics   *237*
A.3   Examples   *240*
         References   *246*

B   **Reinforcement Learning for Control**   *247*
B.1   Markov decision processes   *247*
B.2   Value functions   *248*
B.3   Iterations   *250*
B.4   TD learning   *251*
         Reference   *258*

Index   *259*

Author Biographies

Wen Yu received the B.S. degree in automatic control from Tsinghua University, Beijing, China in 1990 and the M.S. and Ph.D. degrees, both in Electrical Engineering, from Northeastern University, Shenyang, China, in 1992 and 1995, respectively. From 1995 to 1996, he served as a lecturer in the Department of Automatic Control at Northeastern University, Shenyang, China. Since 1996, he has been with CINVESTAV-IPN (National Polytechnic Institute), Mexico City, Mexico, where he is currently a professor with the Departamento de Control Automatico. From 2002 to 2003, he held research positions with the Instituto Mexicano del Petroleo. He was a Senior Visiting Research Fellow with Queen's University Belfast, Belfast, U.K., from 2006 to 2007, and a Visiting Associate Professor with the University of California, Santa Cruz, from 2009 to 2010. He also holds a visiting professorship at Northeastern University in China from 2006. Dr.Wen Yu serves as associate editors of *IEEE Transactions on Cybernetics, Neurocomputing*, and *Journal of Intelligent and Fuzzy Systems*. He is a member of the Mexican Academy of Sciences.

Adolfo Perrusquía (Member, IEEE) received the B. Eng. degree in Mechatronic Engineering from the Interdisciplinary Professional Unit on Engineering and Advanced Technologies of the National Polytechnic Institute (UPIITA-IPN), Mexico, in 2014; and the M.Sc. and Ph.D. degrees, both in Automatic Control from the Automatic Control Department at the Center for Research and Advanced Studies of the National Polytechnic Institute (CINVESTAV-IPN), Mexico, in 2016 and 2020, respectively. He is currently a research fellow in Cranfield University. He is a member of the IEEE Computational Intelligence Society. His main research of interest focuses on robotics, mechanisms, machine learning, reinforcement learning, nonlinear control, system modelling, and system identification.

List of Figures

Figure 1.1 Classic robot control. *4*

Figure 1.2 Model compensation control. *4*

Figure 1.3 Position/force control. *5*

Figure 1.4 Reinforcement learning for control. *7*

Figure 2.1 RLC circuit. *18*

Figure 2.2 Mass-spring-damper system. *18*

Figure 2.3 Position control. *20*

Figure 2.4 Force control. *20*

Figure 2.5 Second-order system for environment and robot. *22*

Figure 2.6 Estimation of damping, stiffness and force. *29*

Figure 3.1 Impedance and admittance control. *39*

Figure 3.2 High stiffness environment in task-space. *41*

Figure 3.3 High stiffness environment in joint space. *42*

Figure 3.4 Low stiffness environment in task space. *43*

Figure 3.5 Low stiffness environment in joint space. *44*

Figure 4.1 Task-space control using joint-space dynamics. *53*

Figure 4.2 Model-free control in high stiffness environment. *56*

Figure 4.3 Position tracking in high stiffness environment. *57*

Figure 4.4 Model-free control in low stiffness environment. *58*

Figure 4.5 Position tracking in low stiffness environment. *59*

Figure 4.6 Pan and tilt robot with force sensor. *61*

Figure 4.7 Environment for the pan and tilt robot. *63*

Figure 4.8 Tracking results in Y. *64*

Figure 4.9   Pan and tilt robot tracking control. *65*

Figure 4.10   4-DOF exoskeleton robot with force/torque sensor. *66*

Figure 4.11   Tracking in joint space. *67*

Figure 4.12   Tracking in task space X. *69*

Figure 4.13   Contact force and trajectory tracking. *70*

Figure 5.1   HITL in joint space. *74*

Figure 5.2   HITL in task space. *80*

Figure 5.3   2-DOF pan and tilt robot. *83*

Figure 5.4   4-DOF exoskeleton robot. *83*

Figure 5.5   Control of pan and tilt robot in joint space. *86*

Figure 5.6   Control of pan and tilt robot in task space. *88*

Figure 5.7   Control of 4-DOF exoskeleton robot in joint space. *89*

Figure 5.8   Torques and forces of 4-DOF exoskeleton robot. *91*

Figure 5.9   Control of 4-DOF exoskeleton robot in task space. *93*

Figure 6.1   Position/force control. *100*

Figure 6.2   Robot-environment interaction. *101*

Figure 6.3   Position/force control with $A_e = -5, B_e = -0.005$. *112*

Figure 6.4   Position/force control with $A_e = -4$ and $B_e = -0.002$. *113*

Figure 6.5   Experimental setup. *114*

Figure 6.6   Environment estimation. *115*

Figure 6.7   Experiment results. *116*

Figure 7.1   Hybrid reinforcement learning. *130*

Figure 7.2   Learning curves. *131*

Figure 7.3   Control results. *132*

Figure 7.4   Hybrid RL in unknown environments. *133*

Figure 7.5   Comparisons of different methods. *134*

Figure 7.6   Learning process of RL. *135*

Figure 8.1   Pole position. *157*

Figure 8.2   Mean error of RL methods. *158*

Figure 8.3   2-DOF planar robot. *159*

Figure 8.4 q_1 position regulation. *160*

Figure 8.5   Control actions of the cart-pole balancing system after 10 seconds. *161*

Figure 8.6   Pole position.   *163*

Figure 8.7   Total cumulative reward curve.   *164*

Figure 8.8   Control input.   *164*

Figure 8.9 Q-function learning curves for $\dot{x}_c = \dot{q} = 0$.   *165*

Figure 8.10   ISE comparisons.   *166*

Figure 8.11   Joint position q_1 tracking.   *167*

Figure 8.12   Total cumulative reward.   *168*

Figure 8.13 Q-function learning curves for $\dot{x}_c = \dot{q} = 0$.   *169*

Figure 9.1   Control methods of redundant robots.   *176*

Figure 9.2   One hidden layer feedforward network.   *178*

Figure 9.3   RL control scheme.   *180*

Figure 9.4   Position tracking of simulations.   *185*

Figure 9.5   MARL Learning curve.   *186*

Figure 9.6   Total reward curve.   *187*

Figure 9.7   Position tracking of experiments.   *188*

Figure 10.1   Tracking results.   *221*

Figure 10.2   Mean squared error.   *222*

Figure 10.3   Learning curves.   *223*

Figure 10.4   Convergence of kernel matrices P_k^i and P_k^c.   *224*

Figure 10.5   Tracking results.   *225*

Figure 10.6   Mean squared error.   *226*

Figure 10.7   Learning curves.   *227*

Figure 10.8   Convergence of kernel matrices $\dot{P}_i$ and $\dot{P}_c$.   *228*

Figure A.1   4-DOF exoskeleton robot.   *240*

Figure A.2   2-DOF pan and tilt robot.   *242*

Figure A.3   2-DOF planar robot.   *245*

Figure A.4   Cart-pole system.   *245*

Figure B.1   Control system in the form of Markov decision process.   *248*

List of Tables

Table 4.1   Model-free controllers gains. *55*

Table 4.2   2-DOF pan and tilt robot control gains. *62*

Table 4.3   Control gains for the 4-DOF exoskeleton. *67*

Table 5.1   Controller gains for the pan and tilt robot. *84*

Table 5.2   Controller gains for the exoskeleton. *84*

Table 6.1   Learning parameters. *111*

Table 6.2   Controllers gains. *114*

Table 8.1   Learning parameters DT RL: cart pole system. *155*

Table 8.2   Learning parameters DT RL: 2-DOF robot. *159*

Table 8.3   Learning parameters CT RL: cart pole system. *162*

Table 8.4   Learning parameters CT RL: 2-DOF robot. *167*

Table 9.1   PID control gains. *184*

Table 10.1   Parameters of the neural RL. *225*

Table A.1   Denavit-Hartenberg parameters of the pan and tilt robot. *241*

Table A.2   Kinematic parameters of the exoskeleton. *242*

Table A.3   Denavit-Hartenberg parameters of the exoskeleton. *243*

Table A.4   2-DOF pan and tilt robot kinematic and dynamic parameters. *244*

Preface

Robot control is a topic of interest for the development of control theory and applications. The main theoretical contributions use the linear and non-linear methods, such that the robot is capable of performing some specific tasks. Robot interaction control is a growing topic for research and industrial applications. The main goal of any robot-interaction control scheme is to achieve a desired performance between the robot and the environment with safe and precise movements. The environment can be any material or system exogenous to the robot, e.g., a human. The robot-interaction controller can be designed for position, force, or both.

Recently reinforcement learning techniques have been applied for optimal and robust control, through the use of dynamic programming theory. They do not require system dynamics and are capable of internal and external changes.

From 2013, the authors started to study human-robot interaction control with intelligent techniques, such as neural networks and fuzzy system. In 2016, the authors put more of their attention on how to solve the human-robot interaction with reinforcement learning. After four years of work, they present their results on model-based and model-free impedance and admittance control, in both joint space and task space. The human-in-the-loop control is analyzed. The model-free optimal robot-interaction control and the design of position/force control using reinforcement learning are discussed. Reinforcement learning methods are studied in large discrete-time space and continuous-time space. For the redundant robots control, we use multi-agent reinforcement learning to solve it. The convergence property of the reinforcement learning is analyzed. The robust human-robot interaction control under the worst-case uncertainty is transformed into the $\mathscr{H}_2/\mathscr{H}_\infty$ problem. The optimal controller is designed and realized by reinforcement learning and neural networks.

We assume the readers are familiar with some applications of robot interaction control, using classical and advanced controllers. We will further develop the systematic analysis for system identification, model-based, and model-free robot interaction controllers. The book is aimed at graduate students, as well as the practitioner engineer. The prerequisites for this book are: robot control, nonlinear systems analysis, in particular Lyapunov approach, neural networks, optimization techniques, and machine learning. The book is useful for a large number of researchers and engineers interested in robot and control.

Many people have contributed to shape this book. The first author wants to thank the financial support of CONACYT under Grant CONACyT-A1-S-8216 and CINVESTAV under Grant SEP-CINVESTAV-62 and Grant CNR-CINVESTAV; he also thanks the time and dedication of his wife, Xiaoou. Without her this book would have not been possible. The second author would like to express his sincere gratitude to his advisor Prof. Wen Yu for the continuous support of his Ph.D. study and research and for his patience, motivation, enthusiasm, and immense knowledge. His guidance helped him throughout his research and writing of this book. Also, he would like to thank Prof. Alberto Soria, Prof. Rubén Garrido, Ing. José de Jesús Meza. Last, but not least, the second author thanks the time and dedication of his parents, Adolfo and Graciela. Without them, this book would have not been possible.

Mexico
Wen Yu, Adolfo Perrusquía

Part I

Human-robot Interaction Control

1

Introduction

1.1 Human-Robot Interaction Control

If we know the robot dynamics, we can use them to design model-based controllers (See Figure 1.1). The famous linear controllers are: Proportional-Derivative (PD) [1], linear quadratic regulator (LQR), and Proportional-Integral-Derivative (PID) [2]. They use linear system theory, so the robot dynamics are required to be linearized at some point of operation. The LQR [3–5] control has been used as a basis for the design of reinforcement learning approaches [6].

The classic controllers use complete or partial knowledge of the robot's dynamics. In these cases (without considering disturbances), it is possible to design controllers that guarantee perfect tracking performance. By using the compensation or the pre-compensation techniques, the robot dynamics is canceled and establishes a simpler desired dynamics [7–9]. The control schemes with model compensation or pre-compensation in joint space can be seen in Figure 1.2. Here q_d is the desired reference, q is the robot's joint position, $e = q_d - q$ is the joint error, u_p is the compensator or pre-compensator of the dynamics, u_c is the control coming from the controller, and $\tau = u_p + u_c$ is the control torque. A typical model-compensation control is the proportional-derivative (PD) controller with gravity compensation, which helps to decrease the steady-state error caused by the gravity terms of the robot dynamics.

When we do not have exact knowledge of the dynamics, it is not possible to design the previous controllers. Therefore, we need to use model-free controllers. Some famous controllers are: PID control [10, 11], sliding mode control [2, 12], and neural control [13]. These controllers are tuned according to specific plant under certain conditions (disturbances, friction, parameters). When new conditions arise, the controllers do not display the same behavior, even reaching instability. Model-free controllers perform well for different tasks and are relatively

Human-Robot Interaction Control Using Reinforcement Learning, First Edition. Wen Yu and Adolfo Perrusquía.
© 2022 The Institute of Electrical and Electronics Engineers, Inc. Published 2022 by John Wiley & Sons, Inc.

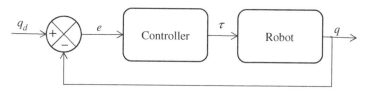

Figure 1.1 Classic robot control

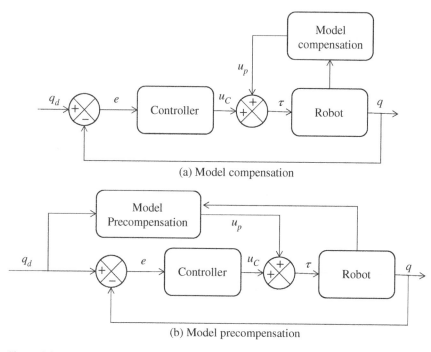

Figure 1.2 Model compensation control

easy to tune; however, they cannot guarantee an optimal performance and require re-tuning the control gains when the robot parameter are changed or a disturbance is applied.

All the above controllers are designed for position control and do not consider interaction with the environment. There is a great diversity of works related to the interaction, such as stiffness control, force control, hybrid control, and impedance control [14]. The force control regulates the interaction force using P (stiffness control), PD, and PID force controllers [15]. The position control can also use force control to perform position and velocity tracking [16, 17] (see Figure 1.3). Here f_d is the desired force, f_e is the contact force, $e_f = f_d - f_e$ is the force error, x_r is the output of the force controller, and $e_r = x_r - x_d$ is the position error in task space.

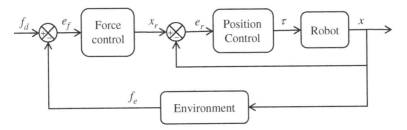

Figure 1.3 Position/force control

The force/position control uses the force for the compensation [17]. It can also use full dynamics to linearize the closed-loop system for perfect tracking [18].

Impedance control [7] addresses the problem of how to move the robot end-effector when it is in contact with the external environment. It uses a desired dynamic model, also known as mechanical impedance, to design the control. The simplest impedance control is the stiffness control, where the stiffness of the robot and the environment have a proportional interaction [19].

Traditional impedance control linearizes the system by assuming that the robot model is known exactly [20–22]. These algorithms need the strong assumption that the exact robot dynamics are known [23]. The robustness of the control lies in the compensation of the model.

Most impedance controllers assume that the desired inertia of the impedance model is equal to the robot inertia. Thus, we only have the stiffness and damping terms, which is equivalent to a PD control law [8, 21, 24]. One way to solve the inaccuracy of dynamic model compensation is through the use of adaptive algorithms, neural networks, or other intelligent methods [9, 25–31]

There are several implementations of impedance control. In [32], the impedance control uses human characteristics to obtain the inertia, damping, and stiffness components of the desired impedance. For the position control a PID control is used, which favors the omission of the model compensation. Another way to avoid the use of the model or to proceed without its full knowledge is to take advantage of system characteristics, that is, the high gear-ratio velocity reduction that causes the non-linear elements to become very small and the system to become decoupled [33].

In mechanical systems, particularly in the haptic field, the admittance is the dynamic mapping from force to motion. The input force "admits" certain amount of movement [11]. The position control based on impedance or admittance needs the inverse impedance model to obtain the reference position [34–38]. This type of scheme is more complete because there is a double control loop where the interaction with the environment can be used more directly.

The applications of impedance/admittance control are quite wide; for example, exoskeletons are used by a human operator. In order to maintain human safety, low mechanical impedance is required, while tracking control requires high impedance to reject the disturbances. So there are different solutions such as frequency molding and the reduction of mechanical impedance using the poles and zeros of the system [39, 40].

Model-based impedance/admittance control is sensitive to modeling error. There exist several modifications to the classical impedance/admittance controllers, such as the position-based impedance control, which improves robustness in the presence of modeling error using an internal position control loop [21].

1.2 Reinforcement Learning for Control

Figure 1.4 shows the control scheme with reinforcement learning. The main difference with the model-free controller in Figure 1.1 is that the reinforcement learning updates its value in each step using the tracking error and control torque.

The reinforcement learning schemes are first designed for discrete-time systems with discrete input space [6, 41]. Among the most famous methods are Monte Carlo [42], Q-learning [43], Sarsa [44], and critic algorithms [45].

If the input space is large or continuous, the classical reinforcement learning algorithms cannot be directly implemented due to the computational cost, and in most cases the algorithm would not converge to a solution [41, 46]. This problem is known as the curse of dimensionality of machine learning. For robot control, the curse of dimensionality increases because there are various degrees of freedom (DOFs), and each DOF needs its own input space [47, 48]. Another factor that makes the dimension problem more acute is the disturbances, because new states and controls must be considered.

To solve the curse of dimensionality, the model-based techniques can be applied to the reinforcement learning [49–51]. These learning methods are very popular; some algorithms are called "policy search" [52–59]. However, these methods require model knowledge to decrease the dimension of the input space.

There are a wide variety of model-free algorithms similar to the discrete-time algorithms. The main idea of these algorithms is to design adequate reward and approximators, which reduces the computational cost in presence of a large or continuous input space.

The simplest approximator to decrease the input space is the handcraft methods [60–65]. They speed up the learning time by looking for regions where the reward is minimized/maximized. [66, 67] use learning methods from input data, similarly to discrete-time learning algorithms, but the learning time increases. Other techniques are based on previously established actions in a sequential and related

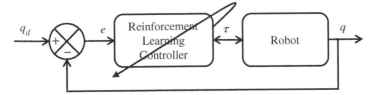

Figure 1.4 Reinforcement learning for control

way; that is, the actions that must be taken at each time instant are defined to do a simple task by themselves [68–72]. The main problems of these methods require an expert knowledge to obtain the best regions and to set the predefined actions.

A linear combination of approximators learn from input data without expert intervention. The most widely used approximators in robot control are inspired by human morphology [73, 74], neural networks [75–77], local models [74, 78], and Gaussian regression processes [79–82]. The success of these approximators is due to the adequate choice of their parameters and hyper-parameters.

A poor reward design can involve a long learning time, convergence to wrong solutions, or the algorithms never converging to any solution. On the other hand, the proper design of a reward helps the algorithm to find the best solutions in each moment of time in a faster way. This problem is known as the "curse of the reward design" [83].

When model-free methods are used, the reward should be designed in such a way that it adapts to changes in the system and possible errors, which is extremely useful in robust control problems where the controller is required to be able to compensate the disturbances or limit disturbances to obtain an optimal performance.

1.3 Structure of the Book

The book consists of two principal parts:

- The first part relates to the design of *human-robot interaction control* in different environments (Chapters 2, 3, 4 and 5).
- The second part deals with *reinforcement learning for robot interaction control* (Chapters 6, 7, 8, 9 and 10).

Part 1

Chapter 2: We address some important concepts for robot interaction control in a mechanical and electrical sense. The concepts of impedance and admittance play important roles for the design of robot interaction control and the environment modeling. The typical environment models and some of the most famous identification techniques for parameters estimation of the environment models are introduced.

Chapter 3: We discuss our first robot-interaction schemes using impedance and admittance controllers. The classical controllers are based on the design of feedback-linearization control laws. The closed-loop dynamics is reduced to a desired dynamics based on the proposed impedance model, which is designed as a second-order linear system. Precision and robustness problems are explained in detail for classical impedance and admittance control. The applicability of these controllers is illustrated by simulations in two different environments.

Chapter 4: We study some model-free controllers that do not need complete knowledge of the robot dynamics. The model-free controllers are designed for an admittance control scheme. The interaction is controlled by the admittance model, while the position controller uses the adaptive control, PID control, or sliding mode control. Stabilities of these controllers are given via Lyapunov stability theory. The applicability of these algorithms is proven via simulations and experiments using different environments and robots.

Chapter 5: We give new robot interaction control scheme known as human-in-the-loop control. Here the environment is the human operator. The human has no contact with the robot. This method uses the input forces/torques of the human operator and maps them into position/orientations of the end-effector via the admittance model. Since the human is in the control loop, she does not know if the applied force/torque yields to singular positions, so it is dangerous for real applications. Therefore, the admittance controllers of the previous chapters are modified to avoid the inverse kinematics, and the Jacobian matrix is modified by using the Euler angles. Experiments illustrate the effectiveness of the approach in both joint and task spaces.

Part 2

Chapter 6: The previous chapters use the desired impedance/admittance model to achieve the desired robot-environment interaction. In most cases, these interactions do not have the optimal performance, they have relative high contact forces or high position errors because they require the environment and robot dynamics. This chapter deals with the reinforcement learning approach for the position/force control in discrete time. The reinforcement learning techniques can achieve a sub-optimal robot-environment interaction.

The optimal impedance model is realized by two different approaches: dynamic programming using a linear quadratic regulator and reinforcement learning. The first one is the model-based control law, the and the second one is model-free. To accelerate the convergence of the reinforcement learning method, the eligibility traces and the temporal difference methods are used. Convergence of the reinforcement learning algorithms is discussed. Stability of the position and force control is analyzed using Lyapunov-like analysis. Simulations and experiments verify the approach in different environments.

Chapter 7: This chapter deals with the large and continuous-time counterpart of the reinforcement learning methods discussed in Chapter 6. Since we are considering big input spaces, the classical reinforcement learning methods cannot handle the problems of the optimal solutions and may not converge. It is required to use approximators to reduce the computational effort and to obtain the reliable optimal or near optimal solutions. This chapter deals with the dimensionality problem in both discrete and continuous time.

We use a parametric approximator based on the normalized radial basis function. The centers of each radial basis function are obtained through the K-means clustering algorithm and random clusters. The convergence of the discrete- and continuous-time versions of the reinforcement learning approximation is analyzed using the contraction property and Lyapunov-like analysis. A hybrid reinforcement learning controller is proposed to take advantage of both discrete- and continuous-time versions. Simulations and experiments are carried out to validate the performance of the algorithms in a position/force control task in different environments.

Chapter 8: We design robust controllers based on a modified reinforcement learning under the worst-case uncertainty, which are robust and present optimal or near optimal solutions. The reinforcement learning methods are designed in both discrete and continuous time. Both methods use a reward designed as an optimization problem under constraints.

In discrete time, the reinforcement learning algorithms are modified using the k-nearest neighbors and the double estimator technique in order to avoid overestimation of the action values. Two algorithms are developed: a large state and discrete action case and a large state-action case. The convergence of the algorithms is analyzed using the contraction property. In continuous time, we use the same algorithms of Chapter 7 under the modified reward. The effectiveness of the robust controllers is proven via simulations and experiments.

Chapter 9: For some kind of robots, such as redundant robots, it is impossible to compute the inverse kinematics or use the Jacobian matrix due to the singularities. This chapter uses the multi-agent reinforcement learning approach to deal with this issue by using only knowledge of the robot forward kinematics. The solutions of the inverse and velocity kinematics of robots and redundant robots are discussed.

To assure controllability and avoid a singularity or multiple solutions, we use the multi-agent reinforcement learning and a proposed double value function method. The kinematic approach is used to avoid the curse of dimensionality. We use small joint displacements as control input until the desired reference is achieved. We discuss the convergence of the algorithm. Simulations and experiments prove the approach with satisfactory results compared with he standard actor-critic methods.

Chapter 10: The reinforcement learning methods that we used in previous chapters learn an optimal control policy from scratch, which translates into large learning time. In order to give previous knowledge to the controller, this chapter gives an $\mathcal{H}_2$ neural control using reinforcement learning in discrete time and continuous time. The controller uses the knowledge of the learned dynamics in order to compute the optimal controller. Convergence of the proposed neural control is analyzed using the contraction property and Lyapunov-like analysis. Simulations are carried out to verify the optimization and robustness of the controller.

Appendices

Appendix A: We discuss some basic concepts and properties of the kinematic and dynamic models of robot manipulators. The dynamics is expressed in both joint space and task space. We also give the kinematic and dynamic models of the robots and systems used in this book by means of the Denavit-Hartenberg convention and the Euler-Lagrange formulation.

Appendix B: We give the basic theory of reinforcement learning and some of the most famous algorithms for controller design. The convergence of the reinforcement learning methods is discussed.

References

1 R. Kelly and V. Santibáñez, *Control de Movimiento de Robots Manipuladores.* Ribera del Loira, España: Pearson Prentice Hall, 2003.

2 A. Perrusquía, W. Yu, A. Soria, and R. Lozano, "Stable admittance control without inverse kinematics," *20th IFAC World Congress (IFAC2017)*, 2017.

3 L. Márton, A. Scottedward Hodel, B. Lantos, and Y. Hung, "Underactuated robot control: Comparing LQR, subspace stabilization, and combined error metric approaches," *IEEE Transactions on Industrial Electronics*, vol. 55, no. 10, pp. 3724–3730, 2008.

4 R. Johansson and M. Spong, "Quadratic optimization of impedance control," *Proc. IEEE International Conference in Robot Autom*, no. 1, pp. 616–621, 1994.

5 M. Matinfar and K. Hashtrudi-Zaad, "Optimization based robot compliance control: geometric and linear quadratic approaches," *Int. J. Robot. Res.*, vol. 24, no. 8, pp. 645–656, 2005.

6 F. Lewis, D. Vrable, and K. Vamvoudakis, "Reinforcement learning and feedback control using natural decision methods to desgin optimal adaptive controllers," *IEEE Control Systems Magazine*, 2012.

7 N. Hogan, "Impedance Control: An Approach to Manipulation," *Journal of Dynamic Systems, Measurement, and Control*, vol. 107, pp. 1–24, March 1985. Transactions of the ASME.

8 F. Ficuciello, L. Villani, and B. Siciliano, "Variable Impedance Control of Redundant Manipulators for Intuitive Human-Robot Physical Interaction," *IEEE Transactions on Robotics*, vol. 31, pp. 850–863, August 2015.

9 A. Khan, D. Yun, M. Ali, J. Han, K. Shin, and C. Han, "Adaptive Impedance Control for Upper Limb Assist Exoskeleton," *IEEE International Conference on Robotics and Automation*, pp. 4359–4366, May 26–30 2015. Seattle, Washington.

10 J. G. Romero, A. Donaire, R. Ortega, and P. Borja, "Global stabilisation of underactuated mechanical systems via PID passivity-based control.," *IFAC-PapersOnLine*, 2017.

11 W. Yu, J. Rosen, and X. Li, "PID Admittance Control for an Upper Limb Exoskeleton," *American Control Conference*, pp. 1124–1129, June 29–July 01 2011. O'Farrel Street, San Francisco, California.

12 R. Xu and U. Özgüner, "Sliding mode control of a class of underactuated systems.," *Automatica*, vol. 44, 2014.

13 Y. Zuo, Y. Wang, X. Liu, S. Yang, L. Huang, X. Wu, and Z. Wang, "Neural network robust H_∞ tracking control strategy for robot manipulators," *Applied Mathematical Modelling*, vol. 34, no. 7, pp. 1823–1838, 2010.

14 S. Chiaverini, B. Siciliano, and L. Villani, "A Survey of Robot Interaction Control Schemes with Experimental Comparison," *IEEE/ASME Transactions on Mechatronics*, vol. 4, pp. 273–285, September 1999.

15 R. Volpe and P. Khosla, "A Theoretical and Experimental Investigation of Explicit Force Control Strategies for Manipulators," *IEEE Transactions on Automatic Control*, vol. 38, pp. 1634–1650, November 1993.

16 M. Heidingsfeld, R. Feuer, K. Karlovic, T. Maier, and O. Sawodny, "A Force-controlled Human-assitive Robot for Laparoscopic Surgery," *IEEE International Conference on Systems, Man, and Cybernetics*, pp. 3435–3439, October 5–8 2014. San Diego, CA, USA.

17 S. Chiaverini, B. Siciliano, and L. Villani, "Force/Position Regulation of Compliant Robot Manipulators," *IEEE Transactions on Automatic Control*, vol. 39, no. 3, pp. 647–652, 1994.

18 O. Khatib, "A Unified Approach for Motion and Force Control of Robot Manipulators: The Operational Space Formulation," *IEEE Journal of Robotics and Automation*, vol. RA-3, pp. 43–53, February 1987.

19 M. Tufail and C. de Silva, "Impedance Control Schemes for Bilateral Teleoperation," *International Conference on Computer Science and Education*, pp. 44–49, August 2014. Vancouver, Canada.

20 F. Caccavele, B. Siciliano, and L. Villani, "The Tricept Robot: Dynamics and Impedance Control," *IEEE/ASME Transactions on Mechatronics*, vol. 8, pp. 263–268, June 2003.

21 S. Kang, M. Jin, and P. Chang, "A Solution to the Accuracy/Robustness Dilemma in Impedance Control," *IEEE/ASME Transactions on Mechatronics*, vol. 14, pp. 282–194, June 2009.

22 R. Bonitz and T. Hsia, "Internal Force-Based Impedance Control for Cooperating Manipulators," *IEEE Transactions on Robotics and Automation*, vol. 12, pp. 78–89, February 1996.

23 R. J. Anderson and M. W. Spong, "Hybrid Impedance Control of Robotic Manipulators," *IEEE Journal of Robotics and Automation*, vol. 4, pp. 549–556, October 1988.

24 T. Tsuji and M. Kaneko, "Noncontact Impedance Control for Redundant Manipulators," *IEEE Transactions on Systems, Man, and Cybernetics*, vol. 29, pp. 184–193, March 1999. Part A: Systems and Humans.

25 S. Singh and D. Popa, "An Analysis of Some Fundamental Problems in Adaptive Control of Force and Impedance Behavior: Theory and Experiments," *IEEE Transactions on Robotics and Automation*, vol. 11, pp. 912–921, December 1995.

26 W. Lu and Q. Meng, "Impedance Control with Adaptation for Robotic Manipulators," *IEEE Transactions on Robotics and Automation*, vol. 7, pp. 408–415, June 1991.

27 A. Abdossalami and S. Sirouspour, "Adaptive Control of Haptic Interaction with Impedance and Admittance Type Virtual Environments," *IEEE*, pp. 145–152, March 13–14 2008. Symposium on Haptic Interfaces for Virtual Environments and Teleoperator Systems.

28 M. Chih and A. Huang, "Adaptive Impedance Control of Robot Manipulators based on Function Approximation Technique," in *Robotica* (Cambridge University Press, ed.), vol. 22, pp. 395–403, 2004.

29 R. Kelly, R. Carelli, M. Amestegui, and R. Ortega, "On Adaptive Impedance Control of Robot Manipulators," *IEEE Robotics and Automation*, vol. 1, pp. 572–577, May 14–19 1989. Scottsdale, AZ.

30 V. Mut, O. Nasisi, R. Carelli, and B. Kuchen, "Tracking Adaptive Impedance Robot Control with Visual Feedback," *IEEE International Conference on Robotics and Automation*, pp. 2002–2007, May 1998. Leuven, Belgium.

31 H. Mohammadi and H. Richter, "Robust Tracking/Impedance Control: Application to Prosthetics," *American Control Conference*, pp. 2673–2678, July 1–3 2015. Chicago, IL, USA.

32 T. Tsuji and Y. Tanaka, "Tracking Control Properties of Human-Robotic Systems Based on Impedance Control," *IEEE Transactions on Systems, Man, and Cybernetics-Part A: Systems and Humans*, vol. 35, pp. 523–535, July 2005.

33 G. Ferreti, G. Magnani, and P. Rocco, "Impedance Control for Elastic Joints Industrial Manipulators," *IEEE Transactions on Robotics and Automation*, vol. 20, pp. 488–498, June 2004.

34 W. Yu, R. Carmona Rodriguez, and X. Li, "Neural PID Admittance Control of a Robot," *American Control Conference*, pp. 4963–4968, June 17–19 2013. Washington, DC, USA.

35 A. Irawan, M. Moktadir, and Y. Tan, "PD-FLC with Admittance Control for Hexapod Robot's Leg Positioning on Seabed," *IEEE American Control Conference*, May 31–June 3, 2015. Kota Kinabalu.

36 K. Tee, R. Yan, and H. Li, "Adaptive Admittance Control of a Robot Manipulator Under Task Space Constraint," *IEEE International Conference on Robotics and Automation*, pp. 5181–5186, May 3–8 2010. Anchorage Convention District, Alaska.

37 W. Yu and J. Rosen, "A Novel Linear PID Controller for an Upper Limb Exoskeleton," *49th IEEE Conference on Decision and Control*, pp. 3548–3553, December 2010. Atlanta, Hotel, GA, USA.

38 M. Dohring and W. Newman, "The Passivity of Natural Admittance Control Implementations," *IEEE International Conference on Robotics and Automation*, pp. 371–376, September 2003. Taipei, Taiwan.

39 S. Oh, H. Woo, and K. Kong, "Frequency-Shaped Impedance Control for Safe Human-Robot Interaction in Reference Tracking Application," *IEEE/ASME Transactions on Mechatronics*, vol. 19, pp. 1907–1916, December 2014.

40 H. Woo and K. Kong, "Controller Design for Mechanical Impedance Reduction," *IEEE/ASME Transactions on Mechatronics*, vol. 20, pp. 845–854, April 2015.

41 L. Buşoniu, R. Babûska, B. De Schutter, and D. Ernst, *Reinforcement learning and dynamic programming using function approximators*. CRC Press, Automation and Control Engineering Series, 2010.

42 R. Sutton and B. A, *Reinforcement Learning: An Introduction*. Cambridge, MA: MIT Press, 1998.

43 H. van Hasselt, "Double Q-learning,"*In Advances in Neural Information Processing Systems (NIPS)*, pp. 2613–2621, 2010.

44 M. Ganger, E. Duryea, and W. Hu, "Double sarsa and double expected sarsa with shallow and deep learning," *Journal of Data Analysis and Information Processing*, 2016.

45 C. Wang, Y. Li, S. Sam Ge, and T. Heng Lee, "Optimal critic learning for robot control in time-varying environments," *IEEE Transactions on Neural Networks and Learning Systems*, vol. 26, no. 10, 2015.

46 W. B. Powell, "AI, OR and control theory: A rosetta stone for stochastic optimization," tech. rep., Princeton University, 2012.

47 S. Hart and R. Grupen, "Learning generalizable control programs," *IEEE Transactions on Autonomous Mental Development*, vol. 3, no. 3, pp. 216–231, 2011.

48 A. G. Barto and S. Mahadevan, "Recent advances in hierarchical reinforcement learning," *Discrete Event Dynamic Systems*, vol. 13, no. 4, pp. 341–379, 2003.

49 C. G. Atkeson, A. Moore, and S. Stefan, "Locally weighted learning for control," *AI Review*, vol. 11, pp. 75–113, 1997.

50 P. Abbeel, A. Coates, M. Quigley, and A. Y. Ng, "An application of reinforcement learning to aerobatic helicopter flight," *In Advances in Neural Information Processing Systems (NIPS)*, 2007.

51 M. P. Deisenroth and C. E. Rasmusen, "PILCO: A model-based and data-efficient approach to policy search," *In 28th International Conference on Machine Learning (ICML)*, 2011.

52 V. Gullapalli, J. Franklin, and H. Benbrahim, "Acquiring robot skills via reinforcement learning," *IEEE Control Systems Magazine*, vol. 14, no. 1, pp. 13–24, 1994.

53 J. Kober and J. Peter, "Policy search for motor primitives in robotics," *In Advances in Neural Information Processing Systems (NIPS)*, 2009.

54 N. Kohl and P. Stone, "Policy gradient reinforcement learning for fast quadrupedal locomotion," *In IEEE International Conference on Robotics and Automation (ICRA)*, 2004.

55 J. A. Bagnell and J. C. Schneider, "Autonomous helicopter control using reinforcement learning policy search methods," *In IEEE International Conference on Robotics and Automation (ICRA)*, 2001.

56 H. Miyamoto, S. Schaal, F. Gandolfo, H. Gomi, Y. Koike, R. Osu, E. Nakano, Y. Wada, and M. Kawato, "A Kendama learning robot bases on bi-directional theory," *Neural Networks*, vol. 9, no. 8, pp. 1281–1302, 1996.

57 J. Peters and S. Schaal, "Learning to control in operational space," *International Journal of Robotics Research*, vol. 27, no. 2, pp. 197–212, 2007.

58 R. Tedrake, T. W. Zhang, and H. S. Seung, "Learning to walk in 20 minutes," *In Yale Workshop on Adaptive and Learning Systems*, 2005.

59 M. P. Deisenroth, C. E. Rasmussen, and D. Fox, "Learning to control low-cost manipulator using data-efficient reinforcement learning.," *In Robotics: Science and Systems (R:SS)*, 2011.

60 H. Benbrahim, J. Doleac, J. Franklin, and O. Selfridge, "Real-time learning: A ball on a beam," *International Joint Conference on Neural Networks (IJCNN)*, 1992.

61 B. Nemec, M. Tamošiūnaitė, F. Wörgötter, and A. Ude, "Task adaption thorough exploration and action sequencing," *IEEE-RAS International Conference on Humanoid Robots (HUMANOIDS)*, 2009.

62 M. Tokic, W. Ertel, and J. Fessler, "The crawler, a class room demonstrator for reinforcement learning," *International Florida Artificial Intelligence Research Society Conference (FLAIRS)*, 2009.

63 H. Kimura, T. Yamashita, and S. Kobayashi, "Reinforcement learning of walking behaviour for a four-legged robot," *IEEE Conference on Decision and Control (CDC)*, 2001.

64 R. A. Willgross and J. Igbal, "Reinforcement learning of behaviors in mobile robots using noisy infrared sensing," *Australian Conference on Robotics and Automation*, 1999.

65 L. Paletta, G. Fritz, F. Kintzler, J. Irran, and G. Dorffner, "Perception and developmental learning of affordances in autonomous robots," *Hertzberg J., Beetz M. and Englert R., editors, KI 2007: Advances in Artificial Intelligence*, vol. 4667, pp. 235–250, 2007. Lecture Notes in Computer Science, Springer.

66 T. Yasuda and K. Ohkura, "A reinforcement learning technique with an adaptive action generator for a multi-robot system," *International Conference on Simulation of Adaptive Behavior (SAB)*, 2008.

67 J. H. Piater, S. Jodogne, R. Detry, D. Kraft, N. Krüger, O. Kroemer, and J. Peters, "Learning visual representations for perception-action systems," *International Journal of Robotics Research*, vol. 30, no. 3, pp. 294–307, 2011.

68 P. Fidelman and P. Stone, "Learning ball acquisition on a physical robot," *International Symposium on Robotics and Automation (ISRA)*, 2004.

69 G. D. Konidaris, S. Kuidersma, R. Grupen, and A. G. Barto, "Autonomous skill acquisition on a mobile manipulator," *AAAI Conference on Artificial Intelligence (AAAI)*, 2011.

70 G. D. Konidars, S. Kuindersma, R. Grupen, and A. G. Barto, "Robot learning from demonstration by constructing skill trees," *International Journal of Robotics Research*, vol. 31, no. 3, pp. 360–375, 2012.

71 R. Platt, R. A. Gruphen, and A. H. Fagg, "Improving grasp skills using schema structured learning," *International Conference on Development and Learning*, 2006.

72 V. Soni and S. P. Singh, "Reinforcement learning of hierarchical skills on the Sony AIBO robot," *International Conference on Development and Learning (ICDL)*, 2006.

73 C. An, C. G. Atkeson, and J. M. Hollerbach, "Model-based control of a robot manipulator," *MIT Press:, Cambridge*, MA, USA, 1988.

74 S. Schaal, "Learning from demonstration," *In Advances in Neural Information Processing Systems (NIPS)*, 1996.

75 Y. Duan, B. Cui, and H. Yang, "Robot navigation based on fuzzy RL algorithm," *International Symposium on Neural Networks (ISNN)*, 2008.

76 C. Gaskett, L. Fletcher, and A. Zelinsky, "Reinforcement learning for a vision based mobile robot," *IEEE/RSJ International Conference on Intelligent Robots and Systems (IROS)*, 2000.

77 R. Hafner and M. Riedmiller, "Reinforcement learning on a omnidirectional mobile robot," *IEEE/RSJ International Conference on Intelligent Robots and Systems (IROS)*, 2003.

78 I. Grondman, M. Vaandrager, L. Buşoniu, R. Babûska, and E. Schuitema, "Efficient model learning methods for actor-critic control," *IEEE Transactions on Systems, man, and cybernetics. Part B: Cybernetics*, vol. 42, no. 3, 2012a.

79 O. Kroemeri, R. Detry, J. Piater, and J. Peters, "Active learning using mean shift optimization for robot grasping," *IEEE/RSJ International Conference on Intelligent Robots and Systems (IROS)*, 2009.

80 O. Kroemer, R. Detry, J. Piater, and J. Peter, "Combining active learning and reactive control for robot grasping," *Robotics and Autonomous Systems*, vol. 58, no. 9, pp. 1105–1116, 2010.

81 A. Rottmann, C. Plagemann, P. Hilgers, and W. Burgard, "Autonomous blimp control using model-free reinforcement learning in a continuous state and action spac," *IEEE/RSJ International Conference on Intelligent Robots and Systems (IROS)*, 2007.

82 K. Gräve, J. Stückler, and S. Behnke, "Learning motion skills from expert demonstrations and own experience using Gaussian process regression," *Joint International Symposium on Robotics (ISR) and German Conference on Robotics (ROBOTIK)*, 2010.

83 B. Kiumarsi, K. G. Vamvoudakis, H. Modares, and F. L. Lewis, "Optimal and autonomous control using reinforcement learning: A survey," *IEEE Transactions on Neural Networks and Learning Systems*, vol. 29, no. 6, 2018.

2

Environment Model of Human-Robot Interaction

2.1 Impedance and Admittance

The impedance concept is commonly used in the electric field; see Figure 2.1, which is a serial RLC circuit. Here V is a voltage source, $I(t)$ is the current of the net, R a resistor, L is an inductor, and C is a capacitor. The impedance is a quotient between the voltage phasor and the current phasor, i.e., it obeys Ohm's law

$$Z = \mathbb{V}/\mathbb{I},$$

where $\mathbb{V}$ is the voltage phasor and $\mathbb{I}$ is the current phasor.

We define Z_R, Z_L, and Z_C as the impedance of the resistor, inductor, and capacitor, respectively. Z_R measures how much the resistor element impedes the flow of charge through the net (the resistors are dissipative elements), Z_L measures how much the inductive element impedes the level of the current through the net (the inductive elements are storage devices), and Z_C measures how much the capacitive element impedes the level of the current through the net (the capacitive elements are storage devices) [1].

The dynamic equation of the RLC circuit in Figure 2.1 is

$$\mathbb{V} = \left(Z_R + Z_L + Z_C \right) \mathbb{I} \tag{2.1}$$

From the above equation we can see the relation between the current and the voltage through the electric impedance. The dynamic equation of the RLC circuit can be also obtained from the Kirchhoff's law as

$$L\frac{dI(t)}{dt} + RI(t) + \frac{1}{C} \int_0^t I(\tau)d\tau = V. \tag{2.2}$$

If we apply the Laplace transform to the differential equation (2.2) with initial conditions zero, it is

$$\left(Ls + R + \frac{1}{Cs} \right) I(s) = V(s), \quad or \quad \left(Ls^2 + Rs + \frac{1}{C} \right) q(s) = V(s), \tag{2.3}$$

where q is the charge.

Human-Robot Interaction Control Using Reinforcement Learning, First Edition. Wen Yu and Adolfo Perrusquía.
© 2022 The Institute of Electrical and Electronics Engineers, Inc. Published 2022 by John Wiley & Sons, Inc.

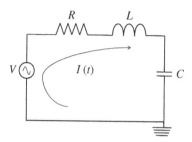

Figure 2.1 RLC circuit.

The first expression shows the impedance relation between the voltage and the circuit current, while the second expression gives a relation between the voltage and the charge, which is also known as impedance filter: $Z(s) = Ls^2 + Rs + 1/C$.

We can write similar relations for mechanical systems. Consider the mass-spring-damper system in Figure 2.2, where m is the car mass, k is the spring stiffness, c is the damping coefficient, and F is the applied force. The car has only horizontal movements, and the spring and damper have linear movements. The dynamic equation that represents the mass-spring-damper system is

$$m\ddot{x} + c\dot{x} + kx = F. \tag{2.4}$$

The Laplace transform of (2.4) is

$$\left(ms^2 + cs + k\right)x(s) = F(s). \tag{2.5}$$

The mechanical impedance or impedance filter is

$$Z(s) = ms^2 + cs + k.$$

Each element has its respective function. The spring stores potential energy (equivalently to a capacitor), the damper dissipates kinetic energy (equivalently to a resistor), and the mass impedes the velocity that the car can obtain (equivalently to an inductor).

The inverse of the impedance is known as admittance. Here the admittance admits a certain amount of movement when a force is applied. For the mechanical

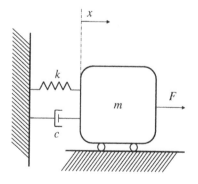

Figure 2.2 Mass-spring-damper system.

system case, the admittance is written as

$$Y(s) = Z^{-1}(s) = \frac{1}{ms^2 + cs + k}. \tag{2.6}$$

The mechanical impedance and admittance models are very useful in environment interaction applications. The environment model is one component of Force control, human-robot interaction, and medical robotics which are examples of an interaction control strategy. If the environment does not change over time, it is usually modeled by a linear spring k and sometimes in parallel with a damper c. Both elements are constants.

For a linear environment, the impedance is defined by the quotient of the Laplace transform of the effort and the flow. In electric systems, the effort is equivalent to the voltage and the flow is the current. In mechanical systems, the effort is a force or torque, and the flow is the linear or angular velocity. For any given frequency ω, the impedance is a complex number with real part $R(\omega)$ and imaginary part $X(\omega)$

$$Z(\omega) = R(\omega) + jX(\omega) \tag{2.7}$$

When ω approaches zero, the magnitude of the environment impedance may have the following possibilities: the magnitude can approach infinity, a finite, nonzero number, or it can approach zero. We introduce the following definitions:

Definition 2.1 *The system with impedance (2.7) is inertial if and only if* $|Z(0)| = 0$.

Definition 2.2 *The system with impedance (2.7) is resistive if and only if* $|Z(0)| = a$ *for some positive* $a \in (0, \infty)$.

Definition 2.3 *The system with impedance (2.7) is capacitive if and only if* $|Z(0)| = \infty$.

The capacitive and inertial environments are dual representations in the sense that the inverse of a capacitive system is inertial, and the inverse of a inertial system is capacitive. A resistive environment is self-dual. In order to represent the duality of the systems we use the equivalent circuits of Norton and Thèvenin [2].

The equivalent circuit of Norton consists of an impedance in parallel with a flow source. The equivalent circuit of Thèvenin consists of an impedance in series with an effort source. The equivalent circuit of Norton represents a capacitive system. The equivalent circuit of Thèvenin presents inertial systems. Both equivalent circuits can be used to represent a resistive system.

A main foundation for control design is that the steady-state error must be zero for a step control input. This can be obtained by considering the following principle.

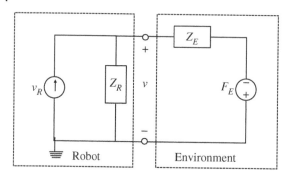

Figure 2.3 Position control.

Duality principle: *The robot manipulator must be controlled such that it can response the dual of the environment.*

This principle can be easily explained by the equivalent circuits of Norton and Thèvenin. When the environment is capacitive, it is represented by an impedance in parallel with a flow source. The robot manipulator (dual) must be represented by an effort source in series with a non-capacitive impedance, either inertial or resistive (see Figure 2.3).

When the environment is inertial, it is represented by an impedance in series with an effort source, and the robot must be represented by a flow source in parallel with a non-inertial impedance, either capacitive or resistive (see Figure 2.4). When the environment is resistive, any equivalent circuit can be used, but the impedance of the manipulator must be no-resistive. In summary, capacitive environments need a robot controlled by force, inertial environments need a robot controlled by position, and resistive environments need both position and force control.

We now show that the duality principle guarantees a zero steady-state error for a step control without inputs in the environment. First it is assumed that the environment is inertial so that

$$Z_E(0) = 0,$$

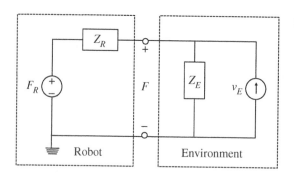

Figure 2.4 Force control.

where the subscript E denotes the environment. Figure 2.3 shows the environment and its respective manipulator, where Z_E is the impedance of the environment, Z_R is the impedance of the robot manipulator, v_R is the input flow, and v is the flow measured between the environment and the robot. The input-output transfer function of the flow is

$$\frac{v}{v_R} = \frac{Z_R(s)}{Z_R(s) + Z_E(s)}. \tag{2.8}$$

If the poles are located at the left semi-plane of the complex plane, the steady-state error for a step control input $1/s$ is given by the final value theorem:

$$e_{ss} = \lim_{t \to \infty} (v - v_R) = \lim_{s \to 0} s(v(s) - v_R(s)) = \frac{-Z_E(0)}{Z_R(0) + Z_E(0)} = 0. \tag{2.9}$$

Here $Z_R(0) \neq 0$, i.e, the manipulator impedance is no-inertial.

We consider that the environment is capacitive so that $Z_E = \infty$. Figure 2.4 shows the environment and its respective dual manipulator, where F_R is the robot effort input. The input/output transfer function for the effort F is

$$\frac{F}{F_R} = \frac{Z_E(s)}{Z_R(s) + Z_E(s)}, \tag{2.10}$$

and its steady-state error for a step input is

$$e_{ss} = \lim_{t \to \infty} (F - F_R) = \lim_{s \to 0} s(F(s) - F_R(s)) = \frac{-Z_R(0)}{Z_R(0) + Z_E(0)} = 0. \tag{2.11}$$

Here $Z_R(0)$ is finite, i.e., the robot impedance is non-capacitive.

For the resistive environment the zero steady-state error is fulfilled, either $Z_R(0) = 0$ and the manipulator is force control, or $Z_R(0) = \infty$ and the manipulator is position controlled.

The principle of duality shows that two different flows or two different efforts cannot be maintained simultaneously at the union port. It is inconsistent to follow the position trajectory and the environment trajectory. Nevertheless, the dual combination of the Norton flow source and a Thèvenin effort source can exist simultaneously [2].

2.2 Impedance Model for Human-Robot Interaction

A robot that is controlled by an impedance is the second-order dynamics [3] that gives a relation between the position of the end-effector and the external force. The features of this relation are governed by the desired impedance values, M_d, B_d, and K_d. They are chosen by the user according to a desired dynamic performance [3].

Here $M_d \in \mathbb{R}^{m \times m}$ is a desired mass matrix, $B_d \in \mathbb{R}^{m \times m}$ is a desired damping matrix, and $K_d \in \mathbb{R}^{m \times m}$ is a desired stiffness matrix. The environment is assumed to be a linear impedance with stiffness and damping parameters (K_e, C_e).

It is usual to treat each Cartesian variable independently, i.e., it is assumed that the environment impedances in different axes are decoupled. Therefore, the desired impedance model is given by the following m differential equations:

$$m_{d_i}(\ddot{x}_i - \ddot{x}_{d_i}) + b_{d_i}(\dot{x}_i - \dot{x}_{d_i}) + k_{d_i}(x_i - x_{d_i}) = f_{e_i}, \tag{2.12}$$

where $i = 1, \cdots, m$, $x_d, \dot{x}_d, \ddot{x}_d \in \mathbb{R}^m$ are the desired position, velocity, and acceleration, and $x, \dot{x}, \ddot{x} \in \mathbb{R}^m$ are the position, velocity and acceleration of the robot end-effector.

In matrix form,

$$M_d(\ddot{x} - \ddot{x}_d) + B_d(\dot{x} - \dot{x}_d) + K_d(x - x_d) = f_e, \tag{2.13}$$

where $f_e = [F_x, F_y, F_z, \tau_x, \tau_y, \tau_x]^T$ is the force/torque vector. The desired impedance matrices are

$$M_d = \begin{bmatrix} m_{d_1} & \cdots & 0 \\ \vdots & \ddots & \vdots \\ 0 & \cdots & m_{d_n} \end{bmatrix}, B_d = \begin{bmatrix} b_{d_1} & \cdots & 0 \\ \vdots & \ddots & \vdots \\ 0 & \cdots & b_{d_n} \end{bmatrix}, K_d = \begin{bmatrix} k_{d_1} & \cdots & 0 \\ \vdots & \ddots & \vdots \\ 0 & \cdots & k_{d_n} \end{bmatrix}. \tag{2.14}$$

The environment impedance matrices are

$$C_e = \begin{bmatrix} c_{e_1} & \cdots & 0 \\ \vdots & \ddots & \vdots \\ 0 & \cdots & c_{e_m} \end{bmatrix}, \quad K_e = \begin{bmatrix} k_{e_1} & \cdots & 0 \\ \vdots & \ddots & \vdots \\ 0 & \cdots & k_{e_m} \end{bmatrix}. \tag{2.15}$$

For 1-DOF robot-environment interaction, the robot and environment have contact (see Figure 2.5). The second-order system is obtained, which is composed by the impedance features of the controller and the environment. If the impedance features of the combined system (impedance controller and environment) can be determined, then the environment features can be computed.

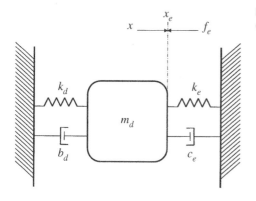

Figure 2.5 Second-order system for environment and robot.

If the environment is located at position x_e, the interaction dynamics of Figure 2.5 is

$$m_d(\ddot{x} - \ddot{x}_d) + b_d(\dot{x} - \dot{x}_d) + k_d(x - x_d) = -c_e\dot{x} + k_e(x_e - x), \tag{2.16}$$

where $x_e \le x$. If $x_e = 0$, the transfer function of the above system is

$$x(s) = x_d(s)\left[1 - \frac{c_e s + k_e}{m_d s^2 + (b_d + c_e)s + k_d + k_e}\right]. \tag{2.17}$$

When there is no interaction, the contact force is zero, and therefore the robot position $x(t)$ is equal to the desired position $x_d(t)$.

The properties of the robot-environment impedance can be determined by its response to a step input. This can be achieved by applying a step input at the robot end-effector and measuring the contact force. For under-damped response, the damped natural frequency ω_d can be determined using a fast Fourier transform (FFT) of the measured contact force. The damping ratio ζ is given by the settling time T_s that gives the convergence time within 5% of the steady-state value:

$$\exp^{-\zeta\omega_n T_s} = 0.05, \tag{2.18}$$

where ω_n is the undamped frequency. The settling time is given by

$$T_s = \frac{2.996}{\zeta\omega_n}, \tag{2.19}$$

and the number of cycles before the settling time is

$$\#\text{cycles} = \frac{2.996\sqrt{1 - \zeta^2}}{2\pi\zeta}. \tag{2.20}$$

The damping ratio is obtained by

$$\zeta = \frac{0.4768}{\sqrt{\#\text{cycles}^2 + 0.2274}}. \tag{2.21}$$

ζ is the number of visible cycles before the contact force converges within 5% of the steady-state error.

The environment stiffness and damping can be obtained by ω_d and ζ. For the robot-environment system as in Figure 2.5, the equivalent stiffness, damping, and mass are

$$k_{eq} = k_d + k_e$$
$$b_{eq} = b_d + c_e \tag{2.22}$$
$$m_{eq} = m_d.$$

In terms of the equivalent values, we can write the natural frequency and damping ratio as

$$\omega_n = \frac{\omega_d}{\sqrt{1 - \zeta^2}} = \sqrt{\frac{k_{eq}}{m_{eq}}}$$

$$\text{and } \zeta = \frac{b_{eq}}{2\sqrt{k_{eq}m_{eq}}}. \tag{2.23}$$

Using the force response of ω_d and ζ, and (2.22)–(2.23), we can compute the environment parameters as:

$$k_e = \omega_n^2 m_d - k_d$$

$$c_e = 2\zeta\sqrt{(k_d + k_e)m_d} - b_d.$$

The main advantage of this method is that it requires few data. Only the contact force has to be measured, which is available in most cases for robot-interaction control schemes. This is an off-line method and does not requires measures of the deflection and velocity of the environment. One disadvantage is that the force response must be under-damped so that we can obtain ω_d and ζ. This implies that the desired impedance values must be chosen carefully to obtain a desired performance. Also this method cannot be easily applied to more complex environments and multiple point contact [4, 5].

2.3   Identification of Human-Robot Interaction Model

We consider the simplest and the most used environment model, the Kelvin-Voigt model, which is a spring-damper system [6], as

$$f_e = C_e\dot{x} + K_e x, \tag{2.24}$$

where $C_e, K_e \in \mathbb{R}^{m \times m}$ are the damping and stiffness matrices of the environment, respectively, $f_e \in \mathbb{R}^m$ is the force/torque vector, and $x, \dot{x} \in \mathbb{R}^n$ are the position and velocity vectors of the robot end-effector, respectively.

There exist several methods for parameter identification. Let us define the regression matrix $\phi = \phi(x, \dot{x}) = \begin{bmatrix} \dot{x} & x \end{bmatrix}^\mathsf{T} \in \mathbb{R}^{p \times m}$ and the parameters vector $\theta = \begin{bmatrix} C_e & K_e \end{bmatrix}^\mathsf{T} \in \mathbb{R}^p$ such that the following linear representation is satisfied

$$f_e = \phi^\mathsf{T}\theta, \tag{2.25}$$

where p is the number of parameters ($p \leq 2m$). The discrete version of (2.25) is

$$F_{e_k} = \phi_k^\mathsf{T}\theta, \tag{2.26}$$

where k is the rate of time step. Let T represent the sampling time interval; then $f_{e_k} = f_e(t_k)$ at time $t = kT$.

The estimated model of (2.26) is

$$\hat{f}_{ek} = \phi_k^{\mathsf{T}} \hat{\theta}, \tag{2.27}$$

where $\hat{f}_{ek} \in \mathbb{R}^m$ is the estimation of the force/torque vector f_{e_k}. $\hat{\theta} \in \mathbb{R}^p$ is the estimation of the parameters vector θ. Here we want to find the parameters vector $\hat{\theta}$ that minimizes the error

$$\tilde{f}_{ek} = \hat{f}_{ek} - f_{e_k} = \phi_k^{\mathsf{T}} \tilde{\theta},$$

where $\tilde{\theta} = \hat{\theta} - \theta$ is the parametric error.

Least-squares method

Let introduce the following index

$$J_1 = \sum_{k=1}^{n} \tilde{f}_{ek}^{\mathsf{T}} \tilde{f}_{ek}.$$

The minimum of the cost index is a zero of the gradient, i.e., $\frac{\partial J_1}{\partial \hat{\theta}} = 0$,

$$\frac{\partial J_1}{\partial \hat{\theta}} = \frac{\partial J_1}{\partial f_{e_k}} \frac{\partial \tilde{f}_{ek}}{\partial \hat{\theta}} = 2 \sum_{k=1}^{n} \tilde{f}_{ek}^{\mathsf{T}} \frac{\partial}{\partial \hat{\theta}} (\phi_k^{\mathsf{T}} \hat{\theta} - f_{e_k})$$

$$= 2 \sum_{k=1}^{n} (\phi_k^{\mathsf{T}} \hat{\theta} - f_{e_k})^{\mathsf{T}} (\phi_k^{\mathsf{T}}) = -2 \sum_{k=1}^{n} f_{e_k}^{\mathsf{T}} \phi_k^{\mathsf{T}} + 2\hat{\theta}^{\mathsf{T}} \sum_{k=1}^{n} \phi_k \phi_k^{\mathsf{T}} = 0.$$

If the inverse of $\sum_{k=1}^{n} \phi_k \phi_k^{\mathsf{T}}$ exists,

$$\hat{\theta} = \left(\sum_{k=1}^{n} \phi_k \phi_k^{\mathsf{T}} \right)^{-1} \sum_{k=1}^{n} \phi_k f_{e_k}. \tag{2.28}$$

(2.28) is the ordinary least squares (LS) [7] in discrete time. In order to design the continuous-time version, we use the following cost index

$$J_2 = \int_0^t \tilde{f}_e^{\mathsf{T}} \tilde{f}_e d\tau.$$

The gradient of the above cost index is

$$\frac{\partial J_2}{\partial \hat{\theta}} = -\int_0^t (2f_e^{\mathsf{T}} \phi^{\mathsf{T}} - 2\hat{\theta}^{\mathsf{T}} \phi \phi^{\mathsf{T}}) d\tau = 0.$$

If $\int_0^t \phi \phi^{\mathsf{T}} d\tau$ exists,

$$\hat{\theta} = \left[\int_0^t \phi \phi^{\mathsf{T}} d\tau \right]^{-1} \int_0^t \phi f_e d\tau. \tag{2.29}$$

The online version of LS is given by [8–10].

Recursive least squares

Let's define the following matrix

$$G_n = \sum_{k=1}^{n} \phi_k \phi_k^\mathsf{T}.$$

Then from (2.28) we have

$$\hat{\theta}_n = G_n^{-1} \sum_{k=1}^{n} \phi_k f_{e_k}$$

and $\hat{\theta}_{n+1} = G_{n+1}^{-1} \sum_{k=1}^{n+1} \phi_k f_{e_k}.$

We can write the above sum as

$$\sum_{k=1}^{n+1} \phi_k f_{e_k} = \sum_{k=1}^{n} \phi_k f_{e_k} + \phi_{n+1} f_{e_{n+1}}$$

$$= G_n \hat{\theta}_n + \phi_{n+1} f_{e_{n+1}} + \phi_{n+1} \phi_{n+1}^\mathsf{T} \hat{\theta}_n - \phi_{n+1} \phi_{n+1}^\mathsf{T} \hat{\theta}_n$$

$$= \left[G_n + \phi_{n+1} \phi_{n+1}^\mathsf{T} \right] \hat{\theta}_n + \phi_{n+1} \left[f_{e_{n+1}} - \phi_{n+1}^\mathsf{T} \hat{\theta}_n \right]$$

$$= G_{n+1} \hat{\theta}_n + \phi_{n+1} \left[f_{e_{n+1}} - \phi_{n+1}^\mathsf{T} \hat{\theta}_n \right].$$

So

$$\hat{\theta}_{n+1} = G_{n+1}^{-1} \left\{ G_{n+1} \hat{\theta}_n + \phi_{n+1} \left[f_{e_{n+1}} - \phi_{n+1}^\mathsf{T} \hat{\theta}_n \right] \right\}$$

$$= \hat{\theta}_n + G_{n+1}^{-1} \phi_{n+1} \left[f_{e_{n+1}} - \phi_{n+1}^\mathsf{T} \hat{\theta}_n \right],$$

because for any non-singular matrices A, B, C, D,

$$(A + BC^{-1}D)^{-1} = A^{-1} - A^{-1}B(C + DA^{-1}B)^{-1}DA^{-1},$$

so let $A = G_n, B = \phi_{n+1}, C = I$, and $D = \phi_{n+1}^\mathsf{T}$,

$$G_{n+1}^{-1} = \left[G_n + \phi_{n+1} \phi_{n+1}^\mathsf{T} \right]^{-1}$$

$$= G_n^{-1} - G_n^{-1} \phi_{n+1} (I + \phi_{n+1}^\mathsf{T} G_n^{-1} \phi_{n+1})^{-1} \phi_{n+1}^\mathsf{T} G_n^{-1}.$$

Let $P_n = G_n^{-1}$ and $P_{n+1} = G_{n+1}^{-1}$. The recursive least squares (RLS) solution is

$$\hat{\theta}_{n+1} = \hat{\theta}_n + P_{n+1} \left[f_{e_{n+1}} - \phi_{n+1}^\mathsf{T} \hat{\theta}_n \right],$$

$$P_{n+1} = P_n - P_n \phi_{n+1} (I + \phi_{n+1}^\mathsf{T} P_n \phi_{n+1})^{-1} \phi_{n+1}^\mathsf{T} P_n. \tag{2.30}$$

The RLS method (2.30) gives an online parameter update method by choosing a big initial P_0 so that it rapidly converges [11–13].

Continuous-time least squares

The RLS method in continuous time works similarly to (2.30). Let's define the following matrix

$$P^{-1} = \int_0^t \phi\phi^\top d\tau, \quad \text{so that} \quad \frac{d}{dt}P^{-1} = \phi\phi^\top.$$

The LS solution (2.29) is

$$\hat{\theta} = P \int_0^t \phi f_e d\tau$$

$$\dot{\hat{\theta}} = \dot{P} \int_0^t \phi f_e d\tau + P\phi f_e.$$

where $P \in \mathbb{R}^{p \times p}$. Obviously matrix P satisfies $P = \left[\int_0^t \phi\phi^\top d\tau \right]^{-1}$.

Now consider the following useful property

$$PP^{-1} = I$$

$$\frac{d}{dt}(PP^{-1}) = 0$$

$$\dot{P}P^{-1} + P\frac{d}{dt}P^{-1} = 0$$

$$\dot{P} = -P\frac{d}{dt}(P^{-1})P = -P\phi\phi^\top P.$$

So

$$\dot{\tilde{\theta}} = \dot{\hat{\theta}} = -P\phi\phi^\top P \int_0^t \phi f_e d\tau + P\phi f_e$$

$$= -P\phi(\phi^\top\hat{\theta} - f_e) = -P\phi\tilde{f}_e = -P\phi\phi^\top\tilde{\theta}.$$

The continuous-time version of the RLS method is

$$\dot{\tilde{\theta}} = -P\phi\phi^\top\tilde{\theta}$$
$$\dot{P} = -P\phi\phi^\top P, \tag{2.31}$$

or

$$\dot{\hat{\theta}} = -P\phi\tilde{f}_e$$
$$\tilde{f}_e = \phi^\top\hat{\theta} - f_e \tag{2.32}$$
$$\dot{P} = -P\phi\phi^\top P.$$

Both the discrete-time and continuous-time version RLS methods can converge rapidly by choosing a big initial matrix P. There are several variants in [14–16].

Gradient method

The gradient method is also known as adaptive identification [17–19]. This method is used for adaptive control. For discrete-time systems, the parameters are updated according to the gradient descent rule:

$$\hat{\theta}_{n+1} = \hat{\theta}_n - \eta \frac{1}{2} \frac{\partial J_1}{\partial \hat{\theta}_n} = \hat{\theta}_n - \eta \phi_n \tilde{f}_{en}, \tag{2.33}$$

or

$$\tilde{\theta}_{n+1} = \tilde{\theta}_n - \eta \phi_n \phi_n^\mathsf{T} \tilde{\theta}_n, \tag{2.34}$$

where $\eta > 0$ is the learning rate.

For continuous-time systems the identification law can be derived by using a Lyapunov function [20] as

$$V = \frac{1}{2} \tilde{\theta}^\mathsf{T} \Gamma^{-1} \tilde{\theta}, \tag{2.35}$$

where $\Gamma \in \mathbb{R}^{p \times p} > 0$ is the adaptive gain matrix. The update law is chosen as

$$\dot{\hat{\theta}} = -\Gamma \phi \tilde{f}_e = -\Gamma \phi \phi^\mathsf{T} \tilde{\theta} \tag{2.36}$$

The time derivative of the Lyapunov function V is

$$\dot{V} = -\tilde{\theta}^\mathsf{T} \phi \phi^\mathsf{T} \tilde{\theta}. \tag{2.37}$$

It is negative definite in terms of the parametric error $\tilde{\theta}$. We can conclude that the parametric error converges to zero, $\tilde{\theta} \to 0$, when $t \to \infty$, as long as the environment estimates $(\hat{K}_e, \hat{C}_e)$ remain bounded and the data are persistent exciting (PE). To guarantee the convergence of the parameter estimates to their real values we need that the input signal must be rich enough so that it excites the plant modes.

If the regression matrix ϕ is PE, then the matrix P or P_n is not singular, and the estimates will be consistent (for LS and RLS methods). For the GM method, if ϕ is PE, the differential equation (2.36) is uniformly asymptotic stable. The following definition establishes the PE condition [21].

Definition 2.4 *The matrix ϕ is persistently exciting (PE) over the time interval $[t, t + T]$ if there exists constants $\beta_1, \beta_2 > 0, T > 0$ such that for all time t the following holds*

$$\beta_1 I \leq \int_t^{t+T} \phi(\sigma) \phi^\mathsf{T}(\sigma) d\sigma \leq \beta_2 I. \tag{2.38}$$

Therefore, for any parameter identification method it is required that the regression matrix is PE to guarantee parameter convergence.

Example 2.1 Consider the 1-DOF Kelvin-Voigt model

$$c_e \dot{x} + k_e x = F,$$

where $c_e = 8$ Ns/m, $k_e = 16$ N/m, and the position is $x = 0.5 + 0.02\sin(20t)$, and assume that the regression matrix is PE. Here we compare the online and continuous-time RLS and GM methods.

Consider the estimated model

$$\hat{c}_e \dot{x} + \hat{k}_e x = \hat{F},$$

which can be in linear form

$$\phi^T(x, \dot{x})\hat{\theta} = \hat{F},$$

where $\phi = [x, \dot{x}]^T$ and $\hat{\theta} = [\hat{k}_e, \hat{c}_e]^T$. The initial conditions of the parameters vector is $\hat{\theta}(0) = [3, 9]^T$. The initial conditions of the RLS method is $P(0) = 1000I_{2\times2}$. The identification gain of the GM method is $\Gamma = 10I_{2\times2}$.

Figure 2.6 Estimation of damping, stiffness and force.

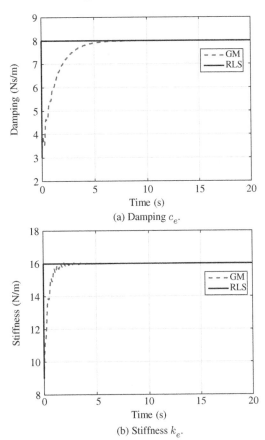

(a) Damping c_e.

(b) Stiffness k_e.

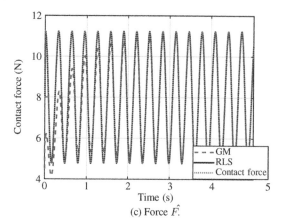

Figure 2.6 (*Continued*)

(c) Force $\hat{F}$.

The results are given in Figure 2.6. The RLS method converges faster than the GM method by choosing a big enough initial condition. If the GM matrix gain is increased, the estimations start to oscillate and never converge. This simulation assumes that there is no noise at the position, velocity, and force measures. It is well known from statistical theory that noisy measures may lead to biased estimates. This problem can be solved by using well-tuned filters.

2.4 Conclusions

In this chapter we address important concepts and properties of the environment in robot interaction control. The impedance and admittance give us a physical (mechanic/electric) approach to understand how the environment parameters affect the behavior of the interaction. The dual principle let us understand the position and force work according to the environment model. Finally, we introduce off-line and online parameter identification methods to obtain the environment parameters.

References

1 R. L. Boylestad, *Introduction to circuit analysis*. Mexico: Pearson Prentice Hall, 2004.

2 R. J. Anderson and M. W. Spong, "Hybrid Impedance Control of Robotic Manipulators," *IEEE Journal of Robotics and Automation*, vol. 4, pp. 549–556, October 1988.

3 N. Hogan, "Impedance Control: An Approach to Manipulation," *Journal of Dynamic Systems, Measurement, and Control*, vol. 107, pp. 1–24, March 1985. Transactions of the ASME.

4 D. Erickson, M. Weber, and I. Sharf, "Contact Stiffness and Damping Estimation for Robotic Systems," *The International Journal of Robotics Research*, vol. 22, pp. 41–57, January 2003.

5 S. Pledgie, K. Barner, S. Agrawal, and T. Rahman, "Tremor Supression Through Impedance Control," *IEEE Transactions on Rehabilitation Engineering*, vol. 8, pp. 53–59, March 2000.

6 T. Yamamoto, M. Bernhardt, A. Peer, M. Buss, and A. M. Okamura, "Techniques for Environment Parameter Estimation During Telemanipulation,"*IEEE/RAS-EMBS International Conference on Biomedical Robotics and Biomechatronics*, pp. 217–223, October 19–22, 2008. Scottsdale, AZ, USA.

7 S. Soliman, A. Al-Kandari, and M. El-Hawari, "Parameter identification method of a separately excited DC motor for speed control," *Electric Machines and Power Systems*, vol. 6, pp. 615–626, 1998.

8 D. Erickson, M. Weber, and I. Sharf, "Contact Stiffness and Damping Estimation for Robotic Systems," *The International Journal of Robotics Research*, vol. 22, pp. 41–57, January 2003.

9 N. Diolaiti, C. Mechiorri, and S. Stramigioli, "Contact Impedance Estimation for Robotic Systems," *IEEE Transactions on Robotics*, vol. 21, pp. 925–935, October 2005.

10 A. Haddadi and K. Hashtrudi-Zaad, "Online Contact Impedance Identification for Robotic Systems," *IEEE/RSJ International Conference on Intelligent Robot and Systems*, pp. 974–980, September 22–26, 2008. Nice, France.

11 A. Janot, P.-O. Vandanjon, and M. Gautier, "Identification of physical parameters and instrumental variables validation with two- stage least squares estimator,"*IEEE Transactions on Control Systems Technology*, vol. 21, no. 4, 2013.

12 M. Gautier, A. Janot, and P.-O. Vandanjon, "A new closed-loop output error method for parameter identification of robot dynamics," *IEEE Transactions on Control Systems Technology*, vol. 21, no. 2, 2013.

13 I. Landau and A. Karimi, "An output error recursive algorithm for unbiased identification in closed loop,"*Automatica*, vol. 33, no. 5, 1997.

14 A. Janot, M. Gautier, A. Jubien, and P. O. Vandanjon, "Comparison between the CLOE method and the DIDIM method for robot identification," *IEEE Transactions on Control Systems Technology*, vol. 22, no. 5, 2014.

15 R. Garrido and A. Concha, "Algebraic identification of a DC servomechanism using Least Squares algorithm," *American Control Conference*, 2011.

16 M. Brunot, A. Janot, F. Carrillo, C. Joono, and J.-P. Noël, "Output error methods for robot identification," *Journal of Dynamic Systems Measurement and Control*, 2019.

17 R. Garrido and R. Miranda, "Closed loop Identification of a DC servomechanism," *IEEE 10th International Conference on Power Electronics CIEP*, 2006.

18 C. Fuh and H. Tsai, "Adaptive parameter identification of servo control systems with noise and high-frequency uncertainties," *Mechanical Systems and Signal Processing*, vol. 21, pp. 1237–1251, 2007.

19 R. Garrido and R. Miranda, "DC servomechanism parameter identification: A closed loop input error approach,"*ISA Transactions*, vol. 51, pp. 42–49, 2012.

20 J. E. Slotine and W. Li, *Applied Nonlinear Control.* Prentice Hall, 1991.

21 S. Sastry and M. Bodson, *Adaptive Control, Stability, Convergence and Robustness.* Prentice Hall, 1989.

3

Model Based Human-Robot Interaction Control

3.1 Task Space Impedance/Admittance Control

The main goal of impedance control is to achieve a desired impedance between the position of the end-effector and the contact force f_e,

$$M_d(\ddot{x} - \ddot{x}_d) + B_d(\dot{x} - \dot{x}_d) + K_d(x - x_d) = f_e, \tag{3.1}$$

where $x, \dot{x}, \ddot{x} \in \mathbb{R}^m$ are the position, velocity, and acceleration of the robot end-effector in Cartesian space, respectively; $x_d, \dot{x}_d, \ddot{x}_d \in \mathbb{R}^m$ are the desired trajectory and its time derivatives; and $M_d, B_d, K_d \in \mathbb{R}^{m \times m}$ are the desired mass, damping, and stiffness of the desired impedance model.

Model-Based Impedance Control

This method was proposed by Hogan [1], and it assumes an exact knowledge of the robot dynamics. It is a feedback linearization control law that establishes a desired dynamic behavior in terms of the desired impedance model.

Consider the joint-space robot dynamics as

$$M(q)J^{-1}(q)\left(\ddot{x} - \dot{J}(q)\dot{q}\right) + C(q, \dot{q})\dot{q} + G(q) = \tau - J^\top(q)f_e. \tag{3.2}$$

Because $\ddot{q} = J^{-1}(q)\left(\ddot{x} - \dot{J}(q)\dot{q}\right)$, the control law of the impedance control is

$$\tau = M(q)J^{-1}(q)\left(u - \dot{J}(q)\dot{q}\right) + C(q, \dot{q})\dot{q} + G(q) + J^\top(q)f_e, \tag{3.3}$$

where

$$u = \ddot{x}_d + M_d^{-1}\left[f_e - B_d(\dot{x} - \dot{x}_d) - K_d(x - x_d)\right]. \tag{3.4}$$

In the Laplace domain,

$$u(s) = M_d^{-1}Z_d(s)x_d(s) - M_d^{-1}f_e(s) - M_d^{-1}(sB_d + K_d)x(s). \tag{3.5}$$

Human-Robot Interaction Control Using Reinforcement Learning, First Edition. Wen Yu and Adolfo Perrusquía.
© 2022 The Institute of Electrical and Electronics Engineers, Inc. Published 2022 by John Wiley & Sons, Inc.

The impedance control is a feedback linearization controller that assumes exact compensation of the robot dynamics. It realizes perfect tracking of the desired impedance model. In most cases, the exact knowledge of the robot dynamics is not available. We can estimate the robot parameters. The control law (3.3) can be written as

$$\tau = \hat{M}(q)J^{-1}(q)\left(u - \dot{J}(q)\dot{q}\right) + \hat{C}(q,\dot{q})\dot{q} + \hat{G}(q) + J^{T}(q)f_e, \qquad (3.6)$$

where $\hat{M}, \hat{C}, \hat{G}$ are estimates of the inertia matrix M, the Coriolis matrix C, and the gravitational torques vector G.

Admittance Control

The desired impedance model is modified by the desired trajectory and the contact force. The trajectory is imposed by position reference in position control loop. The position reference is the solution of the desired impedance model

$$M_d(\ddot{x}_r - \ddot{x}_d) + B_d(\dot{x}_r - \dot{x}_d) + K_d(x_r - x_d) = f_e, \qquad (3.7)$$

where $x_r, \dot{x}_r, \ddot{x}_r \in \mathbb{R}^m$ are the position reference of the inner control loop and its time derivatives, respectively. The command reference is

$$\ddot{x}_r = \ddot{x}_d + M_d^{-1}\left(f_e - B_d(\dot{x}_r - \dot{x}_d) - K_d(x_r - x_d)\right). \qquad (3.8)$$

By the Laplace transform, the desired impedance model is written as

$$M_d s^2(x_r(s) - x_d(s)) + B_d s(x_r(s) - x_d(s)) + K_d(x_r(s) - x_d(s)) = f_e(s),$$
$$x_r(s) = x_d(s) + Z_d^{-1}(s)f_e(s), \qquad (3.9)$$

where

$$Z_d(s) = s^2 M_d + s B_d + K_d.$$

From (3.9) if there is no contact force, the position reference x_r is equal to the desired reference x_d.

Any position control law can be used for the inner control loop. The admittance control uses the same model compensation as in (3.6),

$$u = \ddot{x}_r - K_v(\dot{x} - \dot{x}_r) - K_p(x - x_r), \qquad (3.10)$$

where $K_p, K_v \in \mathbb{R}^{m \times m}$ are the proportional and derivative (PD) diagonal matrices gains, which are designed independently with the desired impedance parameters.

Reference Position

The second-order impedance model (3.7) without force is

$$Z_d(s) = M_d s^2 + B_d s + K_d = 0. \tag{3.11}$$

The characteristic roots of the desired impedance model are

$$s_{1,2}I = \frac{-M_d^{-1}B_d \pm \sqrt{M_d^{-2}B_d^2 - 4M_d^{-1}K_d}}{2}, \tag{3.12}$$

since the matrices M_d, B_d, and K_d are diagonal and positive definite. From (3.9),

$$x_r(s) = x_d(s) + Z_d^{-1}(s)f_e(s) = x_d(s) + \left(M_d s^2 + B_d s + K_d\right)^{-1}f_e(s). \tag{3.13}$$

The desired impedance can be written as the product of the characteristic roots

$$\begin{aligned} x_r(s) &= x_d(s) + \left[(sI - s_1 I)(sI - s_2 I)\right]^{-1}f_e(s) \\ &= x_d(s) + \left[W(sI - s_1 I)^{-1} + V(sI - s_2 I)^{-1}\right]f_e(s), \end{aligned} \tag{3.14}$$

where W and $V = -W$ are matrices that are obtained by solving partial fractions decomposition

$$W = \left(\sqrt{M_d^{-2}B_d^2 - 4M_d^{-1}K_d}\right)^{-1}. \tag{3.15}$$

Therefore, the solution of the position reference is

$$x_r(t) = x_d(t) + W \int_0^t \left(\exp^{s_1(t-\sigma)} - \exp^{s_2(t-\sigma)}\right)f_e(\sigma)d\sigma. \tag{3.16}$$

The characteristic roots are

$$s_{1,2}I = -r \pm p, \tag{3.17}$$

where $r = \frac{M_d^{-1}B_d}{2}$, and $p = \frac{\sqrt{M_d^{-2}B_d^2 - 4M_d^{-1}K_d}}{2}$. Then (3.16) is rewritten as

$$x_r(t) = x_d(t) + W \int_0^t \exp^{-r(t-\sigma)} \left(\exp^{p(t-\sigma)} - \exp^{-p(t-\sigma)}\right)f_e(\sigma)d\sigma. \tag{3.18}$$

Because $\sinh(x) = \frac{e^x - e^{-x}}{2}$, the solution of the position reference is

$$x_r(t) = x_d(t) + 2W \int_0^t \exp^{-r(t-\sigma)} \sinh\left(p(t-\sigma)\right)f_e(\sigma)d\sigma. \tag{3.19}$$

Another way to compute the reference position is to use the state space. Let

$$x_1 = x_r - x_d,$$

$$\dot{x}_1 = \dot{x}_r - \dot{x}_d = x_2, \text{ and}$$

$$\dot{x}_2 = M_d^{-1} \left[f_e - B_d x_2 - K_d x_1 \right].$$

In matrix form,

$$\dot{x} = \begin{bmatrix} \dot{x}_1 \\ \dot{x}_2 \end{bmatrix} = \underbrace{\begin{bmatrix} 0_{m \times m} & I_{m \times m} \\ -M_d^{-1} K_d & -M_d^{-1} B_d \end{bmatrix}}_{A} \underbrace{\begin{bmatrix} x_1 \\ x_2 \end{bmatrix}}_{x} + \underbrace{\begin{bmatrix} 0_{m \times m} \\ M_d^{-1} \end{bmatrix}}_{B} f_e. \tag{3.20}$$

The system above is a linear system, and its solution with zero initial condition is

$$x(t) = \int_0^t \exp^{A(t-\sigma)} B f_e(\sigma) d\sigma. \tag{3.21}$$

The main problem of (3.21) is its lack of generality. For this reason, it cannot be applied.

3.2 Joint Space Impedance Control

In joint space, the inverse kinematics (A.2) is

$$q_r = invf(x_r), \tag{3.22}$$

where $q_r \in \mathbb{R}^n$ is the joint space nominal reference. Then the impedance control in joint space is

$$\tau = M(q)u + C(q, \dot{q})\dot{q} + G(q) + J^T(q)f_e, \tag{3.23}$$

where u is the same as in (3.4). The classical admittance control law in joint space is (3.23), and u is modified as

$$u = \ddot{q}_r - K_p(q - q_r) - K_v(\dot{q} - \dot{q}_r), \tag{3.24}$$

where $\ddot{q}_r \in \mathbb{R}^n$ is the acceleration of the joint space reference. Here the dimensions of proportional and derivative gains are changed from $m \times m$ to $n \times n$.

Joint space controllers are easier than the task space controllers since we avoid the use of the Jacobian and the task space dynamics. However, we have to use the inverse kinematics, which may have different solutions in task space. Also, singularities may lead to instability of the impedance model.

For robots with many DOFs such as redundant robots, the inverse kinematics is local and we cannot obtain a reliable solution for all DOFs. If the robot has few DOFs and the inverse kinematics solution is available, it is preferable to use joint space control.

3.3 Accuracy and Robustness

An equivalent form of the dynamics (3.2) is

$$\tau - J^{T}(q)f_{e} = \hat{M}(q)J^{-1}(q)\left(\ddot{x} - \dot{J}(q)\dot{q}\right) + \hat{C}(q,\dot{q})\dot{q} + \hat{G}(q)$$

$$+ \underbrace{\left(M(q) - \hat{M}(q)\right)J^{-1}(q)\left(\ddot{x} - \dot{J}(q)\dot{q}\right)}_{-\tilde{M}(q)} + \underbrace{\left(C(q,\dot{q}) - \hat{C}(q,\dot{q})\right)\dot{q}}_{-\tilde{C}(q,\dot{q})}$$

$$+ \underbrace{\left(G(q) - \hat{G}(q)\right)}_{-\tilde{G}(q)}. \tag{3.25}$$

We use the control law (3.6) to the dynamics (3.25),

$$\hat{M}(q)J^{-1}(u - \ddot{x}) = -\tilde{M}(q)J^{-1}(q)\left(\ddot{x} - \dot{J}(q)\dot{q}\right) - \tilde{C}(q,\dot{q})\dot{q} - \tilde{G}(q). \tag{3.26}$$

So (3.26) is a common representation for both the impedance control and the admittance control, where u depends on the controller design. We multiply by $\left(\hat{M}(q)J^{-1}(q)\right)^{-1}$ both sides of (3.26), and define the dynamics estimation error $\eta \in \mathbb{R}^{m}$:

$$\eta \triangleq -J(q)\hat{M}^{-1}(q)\left[\tilde{M}(q)J^{-1}(q)\left(\ddot{x} - \dot{J}(q)\dot{q}\right) + \tilde{C}(q,\dot{q})\dot{q} + \tilde{G}(q)\right]. \tag{3.27}$$

Then

$$\eta = u - \ddot{x}. \tag{3.28}$$

(3.28) shows that u and $\ddot{x}$ have second-order linear dynamics

$$u(s) = s^{2}x(s) + \eta(s). \tag{3.29}$$

The control law (3.5) has two parts

$$u(s) = v(s) - M_{d}^{-1}(sB_{d} + K_{d})x(s), \tag{3.30}$$

where $v \in \mathbb{R}^{m}$ is defined as

$$v(s) \triangleq M_{d}^{-1}Z_{d}(s)x_{d}(s) - M_{d}^{-1}f_{e}(s), \tag{3.31}$$

$v(s)$ represents the combination of the desired input reference and the interaction force, and the other part of u is the position feedback input. Substituting (3.30) in (3.29) yields

$$x(s) = Z_{d}^{-1}(s)M_{d}\left(v(s) - \eta(s)\right), \tag{3.32}$$

and $v(s)$ can be rewritten by substituting (3.9) in (3.31)

$$v(s) = M_{d}^{-1}Z_{d}(s)x_{r}(s). \tag{3.33}$$

Substituting (3.33) in (3.32) yields

$$x_d(s) - x(s) = Z_d^{-1}(s)M_d\eta(s)$$
$$= \left(s^2I + sM_d^{-1}B_d + M_d^{-1}K_d\right)^{-1}\eta(s). \tag{3.34}$$

Because of the modeling error, there are not adjustable parameters to the effect of η except for the desired impedance parameters. The impedance control is sensitive to modeling error. When the stiffness is small and the mass is big, the sensitivity is high.

Similar to (3.30), the control input (3.10) can be divided into two parts:

$$u(s) = w(s) - \left(sK_v + K_p\right)x(s), \tag{3.35}$$

where $w \in \mathbb{R}^m$ is defined as

$$w(s) \triangleq \left(s^2I + sK_v + K_p\right)x_r(s) = C(s)x_r(s). \tag{3.36}$$

The first part of (3.35) is the reference input, and the second part is the feedback input. Substituting (3.35) in (3.29) yields

$$x(s) = C^{-1}(s)\left(w(s) - \eta(s)\right). \tag{3.37}$$

Substituting (3.36) in (3.37) gives

$$x_r(s) - x(s) = C^{-1}(s)\eta(s)$$
$$= \left(s^2I + sK_v + K_p\right)^{-1}\eta(s). \tag{3.38}$$

$C^{-1}(s)$ in (3.34) and (3.38) contains the free gains K_p and K_v, which can attenuate the effect of η. Thus the admittance control is more robust than the impedance control. Nevertheless, the dynamics of $C^{-1}(s)$ is excited by η and hinders accurate realization of the desired impedance. Therefore, the lack of accuracy is related to the inner loop dynamics $C^{-1}(s)$. In the worst case, the robot loses contact and starts to oscillate. The impedance control and admittance control schemes are shown in Figure 3.1.

This impedance control has the accuracy/robustness dilemma problem. It is sensitive to the modeling error. Admittance control enhances robustness against the modeling error. However, the desired impedance is not so good, and its accuracy is related to the inner loop dynamics [2]. It is one of the most important problems of robot-interaction control.

The impedance control and the admittance controller use the joint space model in the feedback linearization control law. The impedance and admittance model are mapped to joint space through the inverse of the Jacobian matrix $J(q)$. If we design the impedance control and admittance control in task space using the principle of virtual work,

$$f_\tau = M_x u + C_x \dot{x} + G_x + f_e, \text{ and}$$
$$\tau = J^T(q)f_\tau, \tag{3.39}$$

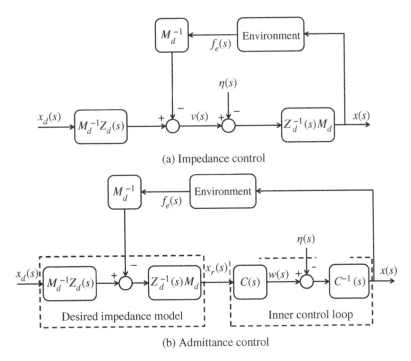

(a) Impedance control

(b) Admittance control

Figure 3.1 Impedance and admittance control.

where u is as in (3.4) or (3.10). The main advantage of this control law is that we avoid the inverse of the Jacobian matrix, so the control law does not present singularities. Nevertheless, (3.39) requires the robot dynamics in task space, which is not very common in robot-interaction tasks.

3.4 Simulations

We use the 4-DOF exoskeleton robot (see Appendix A.1) to show the performance of the impedance control and admittance control schemes. We have complete knowledge of the robot dynamics. The initial conditions of the exoskeleton robot are $q(0) = [0, \pi/2, 0, \pi/4]^\top$ and $\dot{q}(0) = [0, 0, 0, 0]^\top$.

The environment uses the Kelvin-Voigt system with unknown stiffness and damping. The force/torque vector has three force components (F_x, F_y, F_z) and one torque component τ_z. The Jacobian is $J(q) \in \mathbb{R}^{4\times4}$. The desired impedance model has the following desired values $M_d = I_{4\times4}$, $B_d = 20I_{4\times4}$ and $K_d = 100I_{4\times4}$. The desired joint space trajectory is

$$q_{1,d}(t) = 0.5\sin(\omega t),$$
$$q_{2,d}(t) = \frac{\pi}{2} + 0.1\cos(\omega t),$$
$$q_{3,d}(t) = 0.5\sin(\omega t),$$
$$q_{4,d}(t) = \frac{\pi}{3} + \frac{\pi}{4}\cos(\omega t),$$

(3.40)

where $\omega = 2\pi f$ is the angular velocity, $f = 1/T$, and T is the sampling period which is set to $T = 12$ s. The PD control gains of the admittance controllers are: $K_p = 50I_{4\times4}$ and $K_v = 10I_{4\times4}$.

Example 3.1 *High Stiffness Environment*

The environment is at the position $x_e = [0,0,0.59]^T$. The contact force is at the Z-axis. The damping and stiffness of the environment are $c_{e_z} = 100$ Ns/m and $k_{e_z} = 20000$ N/m, respectively. The environment stiffness is bigger than the desired impedance stiffness. The natural frequency and damping ratio of the robot-environment interaction can be obtained from equations (2.3); their values are

$$\omega_n := \sqrt{100 + 20000} = 141.77\,\text{rad/s}, \qquad \zeta = \frac{20 + 100}{2\omega_n} = 0.42,$$

which corresponds to an under-damped behavior.

The tracking results are given in Figure 3.2. The joint positions are shown in Figure 3.3. Since the environment has a higher stiffness than the desired impedance, the position of the end-effector is near the location of the environment, i.e., $x \approx x_e$. The impedance model changes the desired position x_d to the position reference x_r. The impedance model changes the desired position due to the interaction with the environment with new position reference x_r. The joint positions do not follow accurately the desired joint positions when the robot has contact with the environment.

Example 3.2 *Low Stiffness Environment*

The environment is located at the same position as in Example 3.1. The environment parameters are $c_{e_z} = 0.5$ Ns/m and $k_{e_z} = 5$ N/m. The natural frequency and damping ratio of the robot-environment interaction are

$$\omega_n = \sqrt{100 + 5} = 10.25\ \text{rad/s and} \qquad \zeta = \frac{20 + 0.5}{2\omega_n} \approx 1,$$

which corresponds to a damped behavior.

The PD control gains and the desired trajectory are the same as Example 3.1. The tracking results are given in Figure 3.4 and Figure 3.5. The environment stiffness is lower than the desired impedance stiffness, so the robot end-effector position is near the desired position reference, i.e., $x \approx x_d$. The joint positions have good

Figure 3.2 High stiffness environment in task-space.

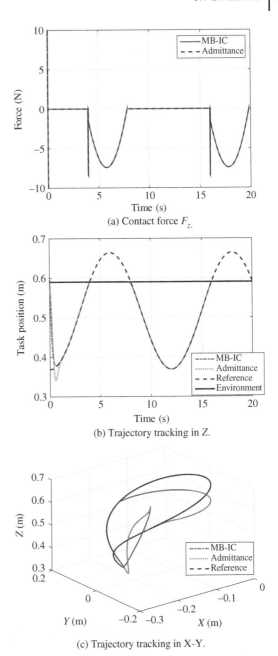

(a) Contact force F_z.

(b) Trajectory tracking in Z.

(c) Trajectory tracking in X-Y.

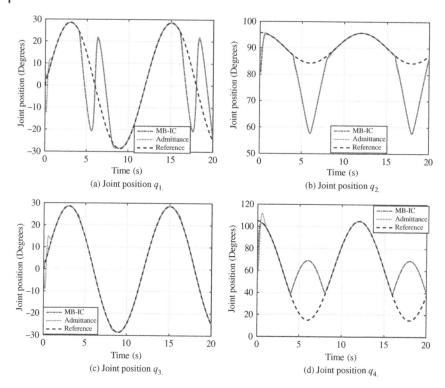

Figure 3.3 High stiffness environment in joint space.

tracking results except for joints q_2 and q_4 because the robot is redundant and the Jacobian does not give a correct mapping from task space to joint space. In order to have a correct mapping we need to use the complete Jacobian.

These two examples show that the impedance and admittance control need exact knowledge of the robot dynamics. This causes the accuracy/robustness dilemma.

3.5 Conclusions

In this chapter we address the classical impedance/admittance controllers in task space and joint space. The response of the robot-environment interaction can be determined according to the environment and desired impedance parameters. We address the accuracy/robustness dilemma of impedance control. A good desired impedance model can be guaranteed when the robot dynamics is known.

Figure 3.4 Low stiffness environment in task space.

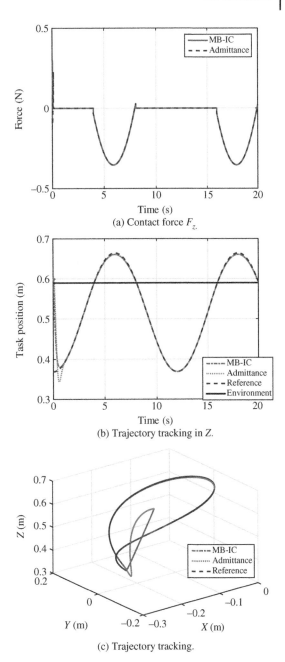

(a) Contact force F_z.

(b) Trajectory tracking in Z.

(c) Trajectory tracking.

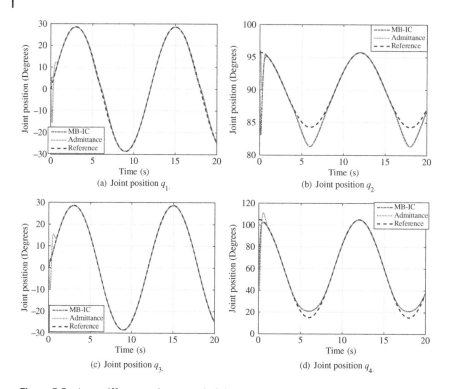

Figure 3.5   Low stiffness environment in joint space.

References

1 N. Hogan, "Impedance Control: An Approach to Manipulation," *Journal of Dynamic Systems, Measurement, and Control*, vol. 107, pp. 1–24, March 1985. Transactions of the ASME.

2 S. Kang, M. Jin, and P. Chang, "A Solution to the Accuracy/Robustness Dilemma in Impedance Control," *IEEE/ASME Transactions on Mechatronics*, vol. 14, pp. 282–194, June 2009.

4

Model Free Human-Robot Interaction Control

4.1 Task-Space Control Using Joint-Space Dynamics

The robot dynamics can be parameterized by the product of a regressor and a parameters vector as

$$M(q)\ddot{q} + C(q, \dot{q})\dot{q} + G(q) = \mathbf{Y}(q, \dot{q}, \ddot{q})\Theta, \tag{4.1}$$

where $\mathbf{Y}(q, \dot{q}, \ddot{q}) \in \mathbb{R}^{n \times p}$ is the regressor that contains all the nonlinear functions of the robot dynamics, and $\Theta \in \mathbb{R}^p$ is the parameters vector.

The model-free admittance control scheme needs position reference, which is obtained from the desired impedance model,

$$M_d \left(\ddot{x}_r - \ddot{x}_d \right) + B_d \left(\dot{x}_r - \dot{x}_d \right) + K_d \left(x_r - x_d \right) = f_e$$
$$\ddot{x}_r = \ddot{x}_d + M_d^{-1} \left[f_e - B_d \left(\dot{x}_r - \dot{x}_d \right) - K_d \left(x_r - x_d \right) \right]. \tag{4.2}$$

The model-free controllers are designed in task space, and they do not require knowledge of the robot inverse kinematics. From (A.5) we have

$$\dot{q} = J^{-1}(q)\dot{x}. \tag{4.3}$$

$\dot{q}_s$ is the nominal reference:

$$\dot{q}_s = J^{-1}(q)\dot{x}_s, \tag{4.4}$$

where $\dot{x}_s$ is the Cartesian nominal reference. From (4.4) and (4.3) the nominal error reference is

$$\Omega = J^{-1}(q) \left(\dot{x} - \dot{x}_s \right), \tag{4.5}$$

where $\dot{x}_s$ depends on the controller design. The robot dynamics is similar to (4.1):

$$M(q)\ddot{q}_s + C(q, \dot{q})q_s + G(q) = \mathbf{Y}_s(q, \dot{q}, \dot{q}_s, \ddot{q}_s)\Theta. \tag{4.6}$$

Human-Robot Interaction Control Using Reinforcement Learning, First Edition. Wen Yu and Adolfo Perrusquía.
© 2022 The Institute of Electrical and Electronics Engineers, Inc. Published 2022 by John Wiley & Sons, Inc.

The position reference x_r has input-state stability [1]:

$$\|x_r\| \leq \|x_d\| + 2\|W\| \int_0^t \exp^{-\lambda_{\max}(r)(t-\sigma)} \sinh(\lambda_{\max}(p)(t-\sigma))\|f_e(\sigma)\| d\sigma$$

$$\leq \|x_d\| + \frac{2\lambda_{\max}(W)\lambda_{\max}(p)}{\lambda_{\max}^2(r) - \lambda_{\max}^2(p)} \sup_{0 \leq \sigma \leq t} \|f_e(\sigma)\|, \tag{4.7}$$

where $r > p$. Because the matrices M_d, B_d, K_d are positive definite and x_r is bounded,

$$\|\dot{x}_r\| \leq \|\dot{x}_d\| + \frac{2\lambda_{\max}(W)\lambda_{\max}(p)}{\lambda_{\max}^2(r) - \lambda_{\max}^2(p)} \sup_{0 \leq \sigma \leq t} \|f_e(\sigma)\|$$

$$\|\ddot{x}_r\| \leq \|\ddot{x}_d\| + \frac{2\lambda_{\max}(W)\lambda_{\max}(p)}{\lambda_{\max}^2(r) - \lambda_{\max}^2(p)} \sup_{0 \leq \sigma \leq t} \|f_e(\sigma)\|, \tag{4.8}$$

$\|f_e\|$ is bounded according to the environment and the desired impedance model.

Admittance Control with Adaptive Gravity Compensation

Proportional-Derivative (PD) control is one of the simplest control techniques for admittance control schemes. Nevertheless, the controller cannot guarantee asymptotic stability when the robot presents gravitational torques, friction, or disturbances.

The classical admittance controllers need complete knowledge of the robot dynamics; otherwise, the admittance controller will have precision and robustness problems. The robot dynamics can be rewritten as

$$\hat{M}(q)\ddot{q}_s + \hat{C}(q,\dot{q})q_s + \hat{G}(q) = Y_s(q,\dot{q},\dot{q}_s,\ddot{q}_s)\hat{\Theta}, \tag{4.9}$$

where $\hat{M}, \hat{C}, \hat{G}$ are estimates of the inertia matrix, the Coriolis matrix, and the gravitational torques vector.

Consider the control law

$$\tau = u + Y_s(q,\dot{q},\dot{q}_s,\ddot{q}_s)\hat{\Theta}$$

and substitute it to (A.17):

$$M(q)\dot{\Omega} + C(q,\dot{q})\Omega = u - J^T(q)f_e + \tilde{M}(q)\ddot{q}_s + \tilde{C}(q,\dot{q})\dot{q}_s + \tilde{G}(q)$$

$$= u - J^T(q)f_e + Y_s(q,\dot{q},\dot{q}_s,\ddot{q}_s)\tilde{\Theta}, \tag{4.10}$$

where $\tilde{M} = \hat{M} - M$, $\tilde{C} = \hat{C} - C$, $\tilde{G} = \hat{G} - G$, and $\tilde{\Theta} = \hat{\Theta} - \Theta$.

Here we want to design an adaptive control law that compensates the gravitational term. The above control law is modified as

$$M(q)\dot{\Omega} + C(q,\dot{q})\Omega = u - J^T(q)f_e + \tilde{G}(q) - M(q)\ddot{q}_s - C(q,\dot{q})\dot{q}_s$$

$$= u - J^T(q)f_e + Y_{s_1}(q)\tilde{\Theta} - M(q)\ddot{q}_s - C(q,\dot{q})\dot{q}_s$$

$$= u - J^T(q)f_e + Y_{s_1}(q)\tilde{\Theta} - Y_{s_2}(q,\dot{q},\dot{q}_s,\ddot{q}_s)\Theta_{s_2}, \tag{4.11}$$

where $\mathbf{Y}_{s_1}(q)\widetilde{\Theta}$ is the estimation error of the gravitational vector and $Y_{s_2}(\cdot)\Theta_{s_2}$ is the parametrization of the inertia and Coriolis matrices in terms of the nominal reference q_s.

The control law is modified as

$$u = J^{\top}(q)f_e - K_s\Omega, \tag{4.12}$$

where $K_s \in \mathbb{R}^{n\times n}$ is a diagonal matrix gain. The nominal Cartesian reference $\dot{x}_s$ is

$$\dot{x}_s = \dot{x}_r - \Lambda\Delta x, \tag{4.13}$$

where $\Lambda \in \mathbb{R}^{m\times m}$ is a diagonal matrix gain and $\Delta x = x - x_r$ is the Cartesian position error. The Cartesian position error is between the end-effector position and the position reference of the desired impedance model.

The final control law is

$$\tau = J^{\top}(q)f_e - K_s J^{-1}(q)(\Delta\dot{x} + \Lambda\Delta x) + \mathbf{Y}_{s_1}(q)\widetilde{\Theta}, \tag{4.14}$$

where $\Delta\dot{x} = \dot{x} - \dot{x}_r$ is the Cartesian velocity error. The closed-loop system dynamics is

$$M(q)\dot{\Omega} + \left(C(q,\dot{q}) + K_s\right)\Omega = \mathbf{Y}_{s_1}(q)\widetilde{\Theta} - \mathbf{Y}_{s_2}(q,\dot{q},\dot{q}_s,\ddot{q}_s)\Theta_{s_2}. \tag{4.15}$$

The Cartesian nominal reference is bounded:

$$\|\dot{x}_s\| \le \|\dot{x}_r\| + \lambda_{\max}(\Lambda)\|\Delta x\|$$
$$\|\ddot{x}_s\| \le \|\ddot{x}_r\| + \lambda_{\max}(\Lambda)\|\Delta\dot{x}\| \tag{4.16}$$

Furthermore, from (A.3),

$$\|\dot{q}_s\| \le \|J^{-1}(q)\|\|\dot{x}_s\| \le \rho_2\|\dot{x}_s\| \tag{4.17}$$
$$\|\ddot{q}_s\| \le \|J^{-1}(q)\| \left\{ \|\ddot{x}_s\| + \|\dot{J}(q)\|\|\dot{q}_s\| \right\}$$
$$\le \rho_2 \left\{ \|\ddot{x}_s\| + \rho_1\rho_2\|\dot{x}_s\| \right\}. \tag{4.18}$$

From (A.18),(A.19),(4.17), and (4.18),

$$\mathbf{Y}_{s_2}(q,\dot{q},\dot{q}_s,\ddot{q}_s)\Theta_{s_2} \le \|M(q)\|\|\ddot{q}_s\| + \|C(q,\dot{q})\|\|\dot{q}_s\|$$
$$\le \beta_1\rho_2 \left\{ \|\ddot{x}_s\| + \rho_1\rho_2\|\dot{x}_s\| \right\} + \beta_2\rho_2\|\dot{q}\|\|\dot{x}_s\|$$
$$\le \chi(t), \tag{4.19}$$

where $\chi(t) = f(\ddot{x}_s,\dot{x}_s,\dot{q},\beta_i,\rho_i)$ is a state-dependent function.

Theorem 4.1 *Consider the robot dynamics (A.17) controlled by the adaptive controller (4.14) and the admittance model (3.8). If the parameters are updated by the adaptive law*

$$\dot{\widetilde{\Theta}} = -K_{\Theta}^{-1}\mathbf{Y}_{s_1}^{\top}(q)\Omega, \tag{4.20}$$

where $K_\Theta \in \mathbb{R}^{p \times p}$ is a diagonal matrix gain, then the closed-loop dynamics is semi-global asymptotic stable and the nominal error converges to a bounded set ε_1 as $t \to \infty$. Also the parameters converge to a small set ε_2 as $\Omega \to \varepsilon_1$ when $t \to \infty$.

Proof: Consider the Lyapunov function candidate

$$V(\Omega, \Theta) = \frac{1}{2}\Omega^T M(q)\Omega + \frac{1}{2}\widetilde{\Theta}^T K_\Theta \widetilde{\Theta}, \tag{4.21}$$

where the first term is the robot kinetic energy in terms of the nominal error reference Ω, and the second term is a quadratic function in terms of the parametric error $\widetilde{\Theta}$. The time derivative of (4.21) along the system trajectories (4.15) is

$$\dot{V} = \Omega^T M(q)\dot{\Omega} + \frac{1}{2}\Omega^T \dot{M}(q)\Omega + \widetilde{\Theta}^T K_\Theta \dot{\widetilde{\Theta}}$$

$$= -\Omega^T \left(-\mathbf{Y}_{s_1}(q)\widetilde{\Theta} + \mathbf{Y}_{s_2}\Theta_{s_2} + \left(C(q,\dot{q}) + K_s - \frac{1}{2}\dot{M}(q) \right)\Omega \right) \tag{4.22}$$

$$+ \widetilde{\Theta}^T K_\Theta \dot{\widetilde{\Theta}}. \tag{4.23}$$

Using the dynamic model properties of Appendix A.2, the above equation is simplified to

$$\dot{V} = -\Omega^T K_s \Omega + \widetilde{\Theta}^T \left(K_\Theta \dot{\widetilde{\Theta}} + \mathbf{Y}_{s_1}^T(q)\Omega \right) - \Omega^T \mathbf{Y}_{s_2}(q, \dot{q}, \dot{q}_s, \ddot{q}_s)\Theta_{s_2}. \tag{4.24}$$

Substituting the adaptive law (4.20) in (4.24) yields

$$\dot{V} = -\Omega^T K_s \Omega - \Omega^T \mathbf{Y}_{s_2}(q, \dot{q}, \dot{q}_s, \ddot{q}_s)\Theta_{s_2}. \tag{4.25}$$

Using (4.19) and (4.25) yields the following inequality

$$\dot{V} \leq -\lambda_{\min}(K_s)\|\Omega\|^2 + \|\Omega\|\chi(t). \tag{4.26}$$

The time derivative (4.26) is negative semi-definite if

$$\|\Omega\| \geq \frac{\chi(t)}{\lambda_{\min}(K_s)} \triangleq \varepsilon_1, \tag{4.27}$$

so there exists a large enough gain K_s such that $K_s \geq |\chi(t)\|$ and the nominal error Ω converges to the bounded set $\varepsilon_1 = \frac{\chi(t)}{\lambda_{\min}(K_s)}$ as $t \to \infty$. The estimates $\hat{\Theta}$ remain bounded if the regressor $\mathbf{Y}_{s_1}$ satisfies the PE condition (2.38). Hence, the trajectories of (4.15) are uniform ultimate bounded (UUB) and semi-global asymptotic stable. ∎

PID Admittance Control

PD control with gravity compensation needs prior knowledge of the gravity term, which is not always available. One of the simplest model-free controllers is the PID (Proportional-Integral-Derivative) control.

The PID admittance control has a similar structure as the adaptive controller; the main difference is at the Cartesian nominal reference:

$$\dot{x}_s = \dot{x}_r - \Lambda\Delta x - \xi,$$
$$\dot{\xi} = \Psi\Delta x,$$

(4.28)

where $\Psi \in \mathbb{R}^{m \times m}$ is a diagonal matrix gain. The control law of the PID admittance control is

$$\tau = J^\mathsf{T}(q)f_e - K_s J^{-1}(q)\left(\Delta\dot{x} + \Lambda\Delta x + \Psi\int_0^t \Delta x d\sigma\right).$$

(4.29)

The closed-loop system of the dynamics (A.17) under the PID control law (4.29) is

$$M(q)\dot{\Omega} + \left(C(q,\dot{q}) + K_s\right)\Omega = -\mathbf{Y}_s(q,\dot{q},\dot{q}_s,\ddot{q}_s)\Theta.$$

(4.30)

It includes complete regressor of the joint nominal reference q_s and its time derivatives. The Cartesian nominal reference is bounded by

$$\|\dot{x}_s\| \le \|\dot{x}_r\| + \lambda_{\max}(\Lambda)\|\Delta x\| + \|\xi\|$$
$$\|\ddot{x}_s\| \le \|\ddot{x}_r\| + \lambda_{\max}(\Lambda)\|\Delta\dot{x}\| + \lambda_{\max}(\Psi)\|\Delta x\|.$$

(4.31)

The regressor is bounded by (A.18),(A.19), (A.23), (4.17), and (4.18):

$$\mathbf{Y}_s(q,\dot{q},\dot{q}_s,\ddot{q}_s)\Theta \le \|M(q)\|\|\ddot{q}_s\| + \|C(q,\dot{q})\|\|\dot{q}_s\| + \|G(q)\|$$
$$\le \beta_1\rho_2\left\{\|\ddot{x}_s\| + \rho_1\rho_2\|\dot{x}_s\|\right\} + \beta_2\rho_2\|\dot{q}\|\|\dot{x}_s\| + \beta_3 \le \chi(t).$$

(4.32)

Theorem 4.2 *The robot (A.17) has the PID control law (4.29) and the admittance controller (3.8). The closed-loop dynamics has semi-global asymptotic stability and the nominal reference converges to a bounded set ε_3 as $t \to \infty$.*

Proof: Consider the following Lyapunov function candidate:

$$V(\Omega) = \frac{1}{2}\Omega^\mathsf{T} M(q)\Omega,$$

(4.33)

which corresponds to the kinetic energy of the closed-loop system. The time derivative of V along the system trajectories (4.30) is

$$\dot{V} = \Omega^\mathsf{T} M(q)\dot{\Omega} + \frac{1}{2}\Omega^\mathsf{T}\dot{M}(q)\Omega$$
$$= -\Omega^\mathsf{T}\left(\mathbf{Y}_s(q,\dot{q},\dot{q}_s,\ddot{q}_s)\Theta + \left(C(q,\dot{q}) + K_s - \frac{1}{2}\dot{M}(q)\right)\Omega\right).$$

(4.34)

Using the robot properties of Appendix A.2,

$$\dot{V} = -\Omega^\mathsf{T} K_s\Omega - \Omega^\mathsf{T}\mathbf{Y}_s(q,\dot{q},\dot{q}_s,\ddot{q}_s)\Theta.$$

(4.35)

Using (4.32) in (4.35),

$$\dot{V} \le -\lambda_{\min}(K_s)\|\Omega\|^2 + \|\Omega\|\chi(t).$$

(4.36)

The time derivative of V is negative definite if the nominal error Ω satisfies

$$\|\Omega\| \geq \frac{\chi(t)}{\lambda_{\min}(K_s)} \triangleq \mu. \tag{4.37}$$

The above condition implies that there exists a large enough gain K_s such that $K_s \geq \|\chi(t)\|$. Moreover, we have the following:

$$\frac{1}{2}\lambda_{\min}(M(q))\|\Omega\|^2 \leq V(\Omega) \leq \frac{1}{2}\lambda_{\max}(M(q))\|\Omega\|^2, \tag{4.38}$$

and Let's define the functions

$$\varphi_1(r) = \frac{1}{2}\lambda_{\min}(M(q))r^2 \text{ and } \varphi_2(r) = \frac{1}{2}\lambda_{\max}(M(q))r^2.$$

The ultimate bound is

$$b = \varphi_1^{-1}\left(\varphi_2(\mu)\right) = \frac{\chi(t)}{\lambda_{\min}(K_s)}\sqrt{\frac{\lambda_{\max}(M(q))}{\lambda_{\min}(M(q))}} \leq \frac{\chi(t)}{\lambda_{\min}(K_s)}\sqrt{\frac{\beta_1}{\beta_0}}, \tag{4.39}$$

which implies that Ω is bounded by

$$\frac{\chi(t)}{\lambda_{\min}(K_s)} \leq \|\Omega\| \leq \frac{\chi(t)}{\lambda_{\min}(K_s)}\sqrt{\frac{\beta_1}{\beta_0}} \tag{4.40}$$

With the above result, we can conclude that the solutions of Ω are uniformly ultimate bounded. Therefore, if we have a large enough gain K_s such that $K_s \geq \|\chi(t)\|$, then the nominal error Ω converges into a bounded set $\varepsilon_3 = \frac{\chi(t)}{\lambda_{\min}(K_s)}$ as $t \to \infty$. ■

PD and PID control do not require knowledge of the robot dynamics. They improve the robustness of the robot-interaction behaviors. Both controllers guarantee convergence to a bounded zone. The accuracy of the model-free controllers is improved because we have an extra element that can make the bounded zone as small as we want.

The adaptive gravity compensation helps to reduce the effect of the gravitational torques vector while the PD gains reduce the effect of the inertia and Coriolis matrices. The integral term of the PID control law helps to reduce the steady-state error. Since Ω is a function of the tracking error and control gains, we can make the tracking error as small as possible by a proper tuning.

Admittance Control with Sliding Mode Compensation

Consider the following Cartesian nominal reference [2, 3]:

$$\dot{x}_s = \dot{x}_r - \Lambda\Delta x - \xi,$$
$$\dot{\xi} = \Psi\text{sgn}(S_x), \tag{4.41}$$
$$S_x = \Delta\dot{x} + \Lambda\Delta x.$$

S_x is the sliding surface that depends on the Cartesian error and its time derivative. The function $\text{sgn}(x) = \begin{bmatrix} \text{sgn}(x_1) & \text{sgn}(x_2) & \dots & \text{sgn}(x_m) \end{bmatrix}^{\mathsf{T}}$ is a discontinuous function of x. The control law is

$$\tau = J^{\mathsf{T}}(q)f_e - K_s J^{-1}(q)\left(\Delta \dot{x} + \Lambda \Delta x + \Psi \int_0^t \text{sgn}(\Delta \dot{x} + \Lambda \Delta x)d\sigma\right). \tag{4.42}$$

The Cartesian nominal reference is bounded by

$$\|\dot{x}_s\| \leq \|\dot{x}_r\| + \lambda_{\max}(\Lambda)\|\Delta x\| + \|\xi\|, \tag{4.43}$$
$$\|\ddot{x}_s\| \leq \|\ddot{x}_r\| + \lambda_{\max}(\Lambda)\|\Delta \dot{x}\| + \lambda_{\max}(\Psi).$$

The closed-loop system under the control law (4.42) satisfies Theorem 4.2, which guarantees semi-global stability of the admittance controller with sliding mode compensation. In order to achieve global stability consider the following super-twisting sliding mode control:

$$\tau = J^{\mathsf{T}}(q)f_e - K_s \Omega - k_1 \|\Omega\|^{1/2}\text{sgn}(\Omega) + \xi, \tag{4.44}$$
$$\dot{\xi} = -k_2\text{sgn}(\Omega),$$

where k_1, k_2 are the sliding mode gains and $\dot{x}_s$ is designed as in (4.13).

The following theorem gives the stability and finite time convergence of admittance control with sliding mode compensation.

Theorem 4.3 *Consider the robot (A.17) controlled by the sliding mode control (4.44) and the admittance model (3.8). If the sliding mode gains satisfy*

$$k_1 > \bar{k}_x, \quad k_2 > \sqrt{\frac{2}{k_1 - \bar{k}_x}}\frac{(k_1 - \bar{k}_x)(1+p)}{(1-p)}, \tag{4.45}$$

where p is constant, $0 < p < 1$, and $\bar{k}_x$ is the upper bound of $Y_s\Theta$, then the tracking error Ω is stable, and it converges to zero in finite time.

Proof: Consider the Lyapunov function

$$V = \frac{1}{2}\zeta^{\mathsf{T}}P\zeta,$$

where $\zeta = [\|\Omega\|^{1/2}\text{sgn}(\Omega), \xi]^{\mathsf{T}}$, and $P = \frac{1}{2}\begin{bmatrix} 4k_2 + k_1^2 & -k_1 \\ -k_1 & 2 \end{bmatrix}$. It is continuous but is not differentiable at $\Omega = 0$.

$$V = 2k_2\|\Omega\| + \frac{\xi^2}{2} + \frac{1}{2}(k_1\|\Omega\|^{1/2}\text{sgn}(\Omega) - \xi)^2.$$

Since $k_1 > 0, k_2 > 0, V$ is positive definite, and

$$\lambda_{\min}(P)\|\zeta\|^2 \leq V \leq \lambda_{\max}(P)\|\zeta\|^2. \tag{4.46}$$

Here $\|\zeta\|^2 = \|\Omega\| + \|\xi\|^2$. The time derivative of V is

$$\dot{V} = -\frac{1}{\|\Omega\|^{1/2}}(\zeta^T Q_1 \zeta - \|\Omega\|^{1/2} Y_s \Theta Q_2^T \zeta), \tag{4.47}$$

where $Q_1 = \frac{k_1}{2}\begin{bmatrix} 2k_2 + k_1^2 & -k_1 \\ -k_1 & 1 \end{bmatrix}$, and $Q_2 = \begin{bmatrix} 2k_2 + \frac{k_1^2}{2} \\ -\frac{k_1}{2} \end{bmatrix}$. Using the inequality

$$\dot{V} \leq -\frac{1}{\|\Omega\|^{1/2}}\zeta^T Q_3 \zeta, \tag{4.48}$$

where

$$Q_3 = \frac{k_1}{2}\begin{bmatrix} 2k_2 + k_1^2 - \left(\frac{4k_2}{k_1} + k_1\right)\bar{k}_x & -(k_1 + 2\bar{k}_x) \\ -(k_1 + 2\bar{k}_x) & 1 \end{bmatrix} \tag{4.49}$$

with the condition (4.45), and $Q_3 > 0$, $\dot{V}$ is negative definite.
From (4.46)

$$\|\Omega\|^{1/2} \leq \|\zeta\| \leq \frac{V^{1/2}}{\lambda_{min}^{1/2}(P)}. \tag{4.50}$$

So

$$\dot{V} \leq -\frac{1}{\|\Omega\|^{1/2}}\zeta^T Q_3 \zeta \leq -\gamma V^{1/2}, \tag{4.51}$$

where $\gamma = \frac{\lambda_{min}^{1/2}(P)\lambda_{min}(Q_3)}{\lambda_{max}(P)} > 0$. Because the solution of the differential equation
$\dot{y} = -\gamma y^{1/2}$ is

$$y(t) = \left[y(0) - \frac{\gamma}{2}t\right]^2,$$

$y(t)$ converges to zero in finite time and reaches zero after $t = \frac{2}{\gamma}y(0)$. Using the comparison principle for (4.51), when $V(\zeta_0) \leq y(0)$, $V(\zeta(t)) \leq y(t)$. So $V(\zeta(t))$ (or Ω) converges to zero after $T = \frac{2}{\gamma}V^{1/2}(\zeta(0))$. ∎

Figure 4.1 shows the general block diagram of the task-space control using joint-space dynamics.

4.2 Task-Space Control Using Task-Space Dynamics

We can use task space dynamics for the previous controllers with the virtual work principle. Consider the nominal error reference in task space:

$$\Omega_x = \dot{x} - \dot{x}_s. \tag{4.52}$$

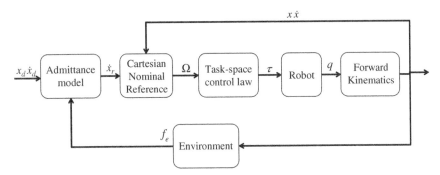

Figure 4.1 Task-space control using joint-space dynamics.

The robot dynamics in task space can be also parameterized in the Cartesian nominal reference:

$$M_x \ddot{x}_s + C_x \dot{x}_s + G_x = Y_x \Theta_x. \tag{4.53}$$

The admittance control with adaptive gravity compensation in the task space domain is

$$f_\tau = f_e - K_s \Omega_x + Y_{x_1} \widetilde{\Theta}_x, \tag{4.54}$$

where $Y_{x_1} \widetilde{\Theta}_x = \widetilde{G}_x$, $K_s \in \mathbb{R}^{m \times m}$, and $\dot{x}_s$ is designed as in (4.13).

If we apply the virtual work principle, the control law does not depend on the gravity component G_x:

$$\tau = J^\top(q)\left(f_e - K_s \Omega_x\right) + Y_{s_1}(q)\widetilde{\Theta}.$$

The task space version of the PID admittance control and the admittance control with sliding mode compensation is

$$f_\tau = f_e - K_s \Omega_x, \tag{4.55}$$

where $\dot{x}_s$ is either as in (4.28) or (4.41). The super-twisting sliding mode control is modified as

$$f_\tau = f_e - K_s \Omega_x - k_1 \|\Omega_x\|^{1/2} \mathrm{sgn}(\Omega_x) + \xi, \\ \dot{\xi} = -k_2 \mathrm{sgn}(\Omega_x). \tag{4.56}$$

The control laws (4.54), (4.55), and (4.56) do not have singularity problems as the joint space dynamics.

4.3 Joint Space Control

Consider the inverse kinematics

$$q_r = invf(x_r).$$

Let's define the joint space nominal error as

$$\Omega_q = \dot{q} - \dot{q}_s.$$

Here $\dot{q}_s$ depends on the controller design. The joint space admittance control with adaptive gravity compensation needs the nominal reference

$$\dot{q}_s = \dot{q}_r - \Lambda \Delta q, \tag{4.57}$$

where Δq is the joint space tracking error, $\dot{q}_r$ is the time-derivative of the joint-space admittance reference q_r, and the dimension of Λ is modified to $n \times n$.

The control law is

$$\tau = J^T f_e - K_s \Omega_q + \mathbf{Y}_{s_1}(q)\widetilde{\Theta}, \tag{4.58}$$

where $K_s \in \mathbb{R}^{n \times n}$.

The joint space control laws for the PID controller and the sliding mode compensation is

$$\tau = J^T(q) f_e - K_s \Omega_q, \tag{4.59}$$

where for the PID control,

$$\dot{q}_s = \dot{q}_r - \Lambda \Delta q - \xi,$$
$$\dot{\xi} = \Psi \Delta q, \tag{4.60}$$

and for the sliding mode compensation,

$$\dot{q}_s = \dot{q}_r - \Lambda \Delta q - \xi,$$
$$\dot{\xi} = \Psi \mathrm{sgn}(S_q), \tag{4.61}$$
$$S_q = \Delta \dot{q} + \Lambda \Delta q,$$

where $\Psi \in \mathbb{R}^{n \times n}$. The super-twisting sliding mode control law is

$$f_\tau = f_e - K_s \Omega_q - k_1 \|\Omega_q\|^{1/2} \mathrm{sgn}(\Omega_q) + \xi,$$
$$\xi = -k_2 \mathrm{sgn}(\Omega_q). \tag{4.62}$$

The main advantage of the joint space control is that we do not have singularity problems. However, the inverse kinematics solution is not always available.

4.4 Simulations

The conditions of the simulations are the same as in Section 3.4. The control gains are given in Table 4.1.

The adaptive law (4.14) only estimates two parameters of the gravity torques vector due to the robot configuration. The Jacobian considers the linear velocity

Table 4.1 Model-free controllers gains.

Gains	Adaptive	PID	Sliding PD
Λ	$10I_{4\times4}$	$15I_{4\times4}$	$10I_{4\times4}$
Ψ	–	$20I_{4\times4}$	$15I_{4\times4}$
K_s	$5I_{4\times4}$	$8I_{4\times4}$	$6I_{4\times4}$
K_Θ	$0.01I_{2\times2}$	–	–

Jacobian and one component of the angular velocity Jacobian in order to obtain a square Jacobian matrix.

Example 4.1 *High and Low Stiffness Environments*
The high and low stiffness environment are the same as in Section 3.4. Figure 4.2 and Figure 4.3 show the high stiffness results. Figure 4.4 and Figure 4.5 show the low stiffness results.

The results show similar performances to the model-free admittance and admittance control in Section 3.4. When the environment stiffness is greater than the desired impedance stiffness, then the admittance control changes the desired position x_d to x_r. It is equivalent to the environment position x_e, i.e., $x_r \approx x_e$. If the environment stiffness is lower than the desired impedance stiffness, the output of the admittance control is almost the same as the desired reference, $x_r \approx x_d$.

The main advantage of these controllers is that they do not need the robot dynamics. The PD control guarantees stability and convergence with steady-state error. The compensation terms, such as adaptive, integral, and sliding mode, reduce the steady-state error such that the end-effector position x converges to the position reference x_r.

Similar to the model-based impedance/admittance controllers, the joints q_2 and q_4 have more errors due to the space transformation, which can be solved by using the complete Jacobian. Nevertheless, the complete Jacobian introduces new singularities to the closed-loop system. The model-free controllers give a reliable solution to the accuracy and robustness dilemma of impedance control.

4.5 Experiments

This section discusses experiments of the model-based and model-free admittance controllers on 2-DOF pan and tilt robot and a 4-DOF exoskeleton robot prototypes with a force/torque sensor mounted at its end effector. The

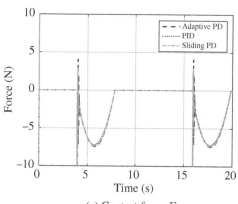

(a) Contact force F_z.

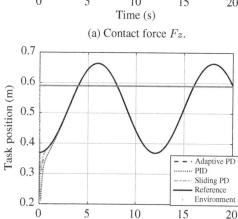

(b) Trajectory tracking in Z.

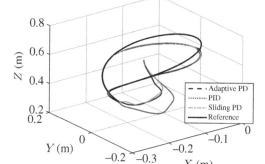

(c) Trajectory tracking in $X - Y - Z$.

Figure 4.2 Model-free control in high stiffness environment.

Figure 4.3 Position tracking in high stiffness environment.

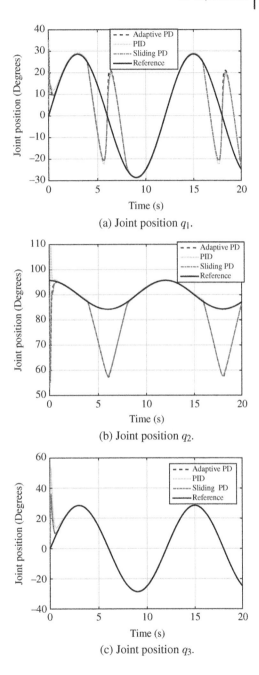

(a) Joint position q_1.

(b) Joint position q_2.

(c) Joint position q_3.

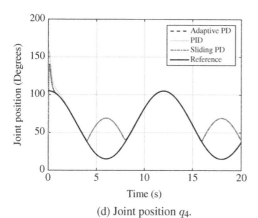

Figure 4.3 (*Continued*)

(d) Joint position q_4.

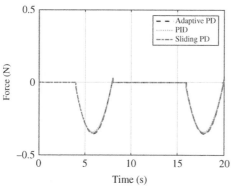

Figure 4.4 Model-free control in low stiffness environment.

(a) Contact force F_z.

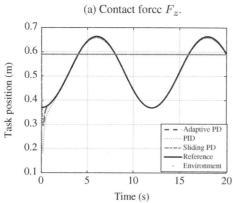

(b) Trajectory tracking in Z.

Figure 4.4 (*Continued*)

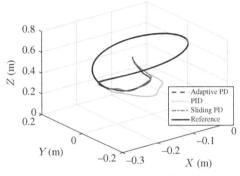

(c) Trajectory tracking in $X - Y - Z$.

Figure 4.5 Position tracking
in low stiffness environment.

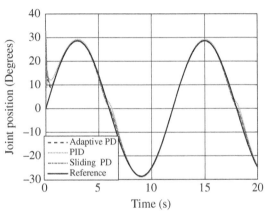

(a) Joint position q_1.

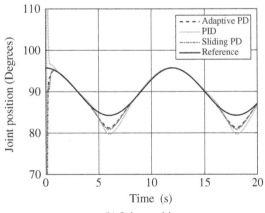

(b) Joint position q_2.

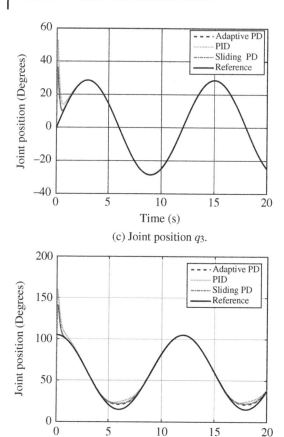

Figure 4.5 (*Continued*)

(c) Joint position q_3.

(d) Joint position q_4.

environment is estimated by the RLS method (3.32). The real-time environment is MATLAB/Simulink®.

The velocity is estimated by the following high-pass filter:

$$v(s) = \frac{bs}{s+b}x(s),$$

where $x(s)$ is the position, $v(s)$ is the estimated velocity, and b is the cutoff frequency of the filter. The experiments use the following high-pass filter:

$$v(s) = \frac{300s}{s+300}x(s).$$

We use the low-pass filter

$$H(s) = \frac{500}{s+500}$$

to smooth the velocity estimates and the force/torque measurements.

The inertia matrix $M(q)$ components are very small, and therefore it reduces the controller effect. The impedance control does not work, because the controller does not have any control gain that makes the control law robust in the presence of modelling error. The impedance control can be implemented by neglecting the inertia matrix component or increasing the desired impedance until it achieves a desired performance. However, this is not convenient. Therefore, we do not present impedance control results.

Example 4.2 *2-DOF Pan and Tilt robot*
Consider the 2-DOF robot pan and tilt robot shown in Appendix A.2. The force/torque sensor is mounted at the robot end effector and is shown in Figure 4.6. The desired joint space trajectory is

$$q_1(t) = \frac{\pi}{8} + \frac{\pi}{8} \cos\left(\frac{\pi}{3}t\right) + 0.001 \sin(40t),$$

$$q_2(t) = -\frac{\pi}{6} + \frac{\pi}{8} \sin\left(\frac{\pi}{3}t\right) + 0.001 \sin(40t).$$

The trajectory is designed to avoid singularities at the upright position of the robot. A small term with high frequency is added as the PE signal for the environment identification. The environment is a wood table with unknown stiffness and damping. The environment is on the Y axis at a distance of 3.4 cm to the robot end

Figure 4.6 Pan and tilt robot with force sensor.

Table 4.2 2-DOF pan and tilt robot control gains.

Gains	Admittance	Adaptive PD	PID	Sliding PD
K_p	$50 \times 10^3 I_{3\times3}$	–	–	–
K_v	$100 I_{3\times3}$	–	–	–
Λ	–	$90 I_{3\times3}$	$90 I_{3\times3}$	$90 I_{3\times3}$
Ψ	–	–	$0.5 I_{3\times3}$	$0.15 I_{3\times3}$
K_Θ	–	1×10^{-3}	–	–
K_s	–	$0.9 I_{2\times2}$	$0.9 I_{2\times2}$	$0.9 I_{2\times2}$

effector. The other axes have free movement. The end-effector is controlled by the task space control with joint space dynamics.

The initial condition of the RLS algorithm is $P(0) = 10000 I_{2\times2}$ with $\hat{\theta}(0) = [0,0]^T$. The desired impedance model is $M_d = I_{3\times3}, B_d = 140 I_{3\times3}, K_d = 4000 I_{3\times3}$. The control gains are shown in Table 4.2.

Here the admittance control gains are big compared with the other controllers because they are tuned with small values, which are from the inertia matrix estimation. The environment estimates and tracking results are given in Figure 4.7, Figure 4.8, and Figure 4.9.

The environment stiffness is approximated as $\hat{k} \approx 480$ N/m, the desired impedance stiffness is $K_d = 4000$ N/m, and $K_d > \hat{k}$, which implies that the robot end effector approximates to the output of the admittance model, i.e., $x \approx x_r$, and because the admittance model modifies the desired reference x_d as the position reference x_r. Both the model-based admittance controller and the model-free controllers have good performance because the robot has high friction and the gravitational terms are neglected.

Example 4.3   4-DOF Exoskeleton Robot
The 4-DOF exoskeleton robot with the force/torque sensor mounted at its end effector is shown in Figure 4.10 and Appendix A.1.

The environment is the same wood table used in the pan and tilt robot experiment. The main difference is that the force/torque sensor is mounted at the X axis instead of the Z axis (see Figure 4.10). This difference implies that the force/torque measurements are not in the same direction as the global reference system. To solve this issue, the following modification is made:

$$F_x, \tau_x \Longrightarrow Y, \beta$$
$$F_y, \tau_y \Longrightarrow Z, \gamma$$
$$F_z, \tau_z \Longrightarrow X, \alpha.$$

Figure 4.10 4-DOF exoskeleton robot with force/torque sensor.

Figure 4.7 Environment for the pan and tilt robot.

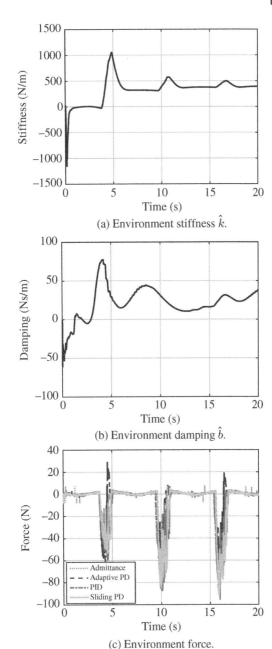

(a) Environment stiffness $\hat{k}$.

(b) Environment damping $\hat{b}$.

(c) Environment force.

ronment is located on the X axis at a distance of 28.5 cm to the robot
tor.

simulation study, the Jacobian matrix is composed of the linear velocity
and one component of the angular velocity Jacobian in order to obtain a
atrix. In this experiment we use two components of the angular velocity
due to sensor mounting. Therefore, $J(q) \in \mathbb{R}^{5 \times 4}$. The task space control
the joint space dynamics is used. The desired joint space trajectory is

$$(t) = 0.3 \sin\left(\frac{\pi}{3}t\right),$$

$$(t) = -0.4 \cos\left(\frac{\pi}{3}t\right),$$

$$(t) = -0.2 - 0.55 \sin\left(\frac{\pi}{3}t\right),$$

$$t) = 0.35 + 0.2 \cos\left(\frac{\pi}{3}t\right).$$

the environment parameters of the pan and tilt robot experiments. The
npedance model is divided in two parts: the first part is for the force
nts whose impedance parameters are $M_d = I_{3\times3}$, $B_d = 140I_{3\times3}$, and $K_d =$
and the second part is for the torque components whose values are $M_d =$
$15I_{2\times2}$, and $K_d = 56I_{2\times2}$. The control gains are given in Table 4.3.
4.11, Figure 4.12, and Figure 4.13 show the tracking results of
:eleton robot. This robot is affected by the gravitational terms, so

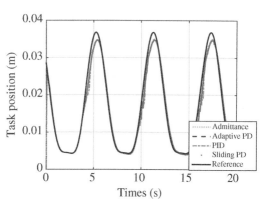

(a) Position reference Y_r.

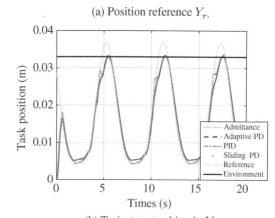

(b) Trajectory tracking in Y.

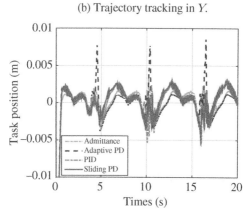

(c) Cartesian position error in Y.

Figure 4.8 Tracking results in Y.

Figure 4.9 Pan and tilt robot tracking control.

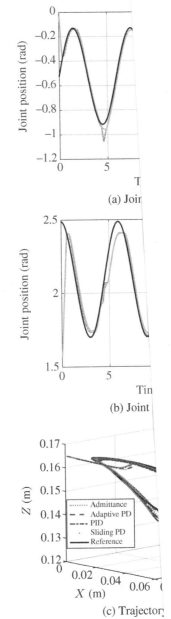

(a) Joint

(b) Joint

(c) Trajectory

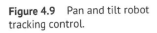

The
end
In
Jacc
squa
Jacc
base

de
co
20
I_2

th

Table 4.3 Control gains for the 4-DOF exoskeleton

Gains	Admittance	Adaptive PD	PID	Sliding PD
K_p	$4 \times 10^3 I_{5\times5}$	–	–	–
K_v	$10I_{5\times5}$	–	–	–
Λ	–	$90I_{5\times5}$	$90I_{5\times5}$	$90I_{5\times5}$
Ψ	–	–	$5I_{5\times5}$	$I_{5\times5}$
K_Θ	–	$0.1I_{2\times2}$	–	–
K_s	–	$3I_{4\times4}$	$2I_{4\times4}$	$2I_{4\times4}$

Figure 4.11 Tracking in joint space.

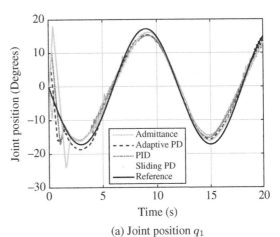

(a) Joint position q_1

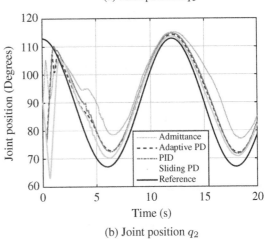

(b) Joint position q_2

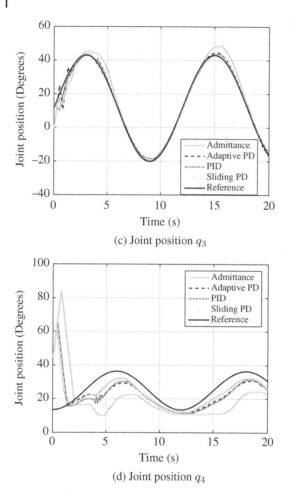

Figure 4.11 *(Continued)*

(c) Joint position q_3

(d) Joint position q_4

the model-based admittance controller loses accuracy. This problem is more evident in the task space error. There is a large tracking error compared with the model-free controllers. This accuracy problem causes the robot to lose contact with the environment. The model-free controllers have good tracking results without the robot dynamics.

4.6 Conclusions

In this chapter we address different model-free controllers in both task space and joint space. These controllers guarantee the position tracking with the admittance

Figure 4.12 Tracking in task space X.

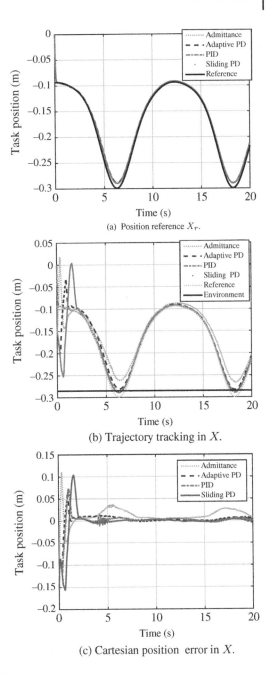

(a) Position reference X_r.

(b) Trajectory tracking in X.

(c) Cartesian position error in X.

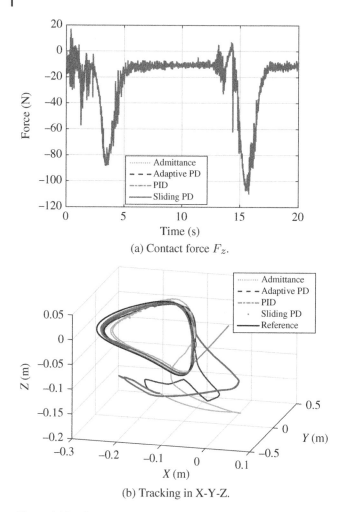

(a) Contact force F_z.

(b) Tracking in X-Y-Z.

Figure 4.13 Contact force and trajectory tracking.

control The accuracy and robustness dilemma of impedance control is solved without the robot dynamics. Three different compensations, adaptive, integral, and sliding mode, reduce the steady-state error. Convergence and stability of the model-free controllers is presented using Lyapunov stability theory. Simulations and experiments verify the effectiveness of the model-free controllers in different environments.

References

1 H. K. Khalil, *Nonlinear Systems*. Upper Saddle River, New Jersey: Prentice Hall, Third ed., 2002.

2 O. A. Dominguez Ramirez, V. Parra Vega, M. G. Diaz Montiel, M. J. Pozas Cardenas, and R. A. Hernandez Gomez, *Cartesian Sliding PD Control of Robot Manipulators for Tracking in Finite Time:Theory and Experiments*, ch. 23, pp. 257–272. Vienna, Austria: DAAAM International, 2008.

3 A. Perrusquía, W. Yu, A. Soria, and R. Lozano, "Stable admittance control without inverse kinematics," *20th IFAC World Congress (IFAC2017)*, 2017.

5

Human-in-the-loop Control Using Euler Angles

5.1 Introduction

The main objective of the human-in-the-loop (HITL) control is to generate specific tasks to combine human skills and robot properties [1], for example, co-manipulation [2], haptic operation [3], [4], and learning from demonstrations [5]. HITL has two loops: the inner loop, which is motion control in joint space, and the outer loop, which is admittance control in task space. HITL is an emerging field in robotics that provides an effective human-robot interaction (HRI) method [6].

In joint space, HITL needs the inverse kinematics to transform the position and orientation of the end effector to the desired positions of the motor so that the movement of the robot is globally decoupled [7, 8]. The inverse kinematics provides multiple solutions for a certain position and orientation and needs a complete knowledge of the kinematics of the manipulator. In task space, HITL needs the Jacobian matrix to transform the control signals to the joint space [9], which also require knowledge of the kinematic parameters.

HITL in both joint space and task space needs a complete model of the robot. The most popular approach for HITL is to use the classical impedance/admittance controllers such that the robot dynamics is compensated and the closed-loop system has the desired impedance model [10]. We demonstrated that the model-free controllers have good results and do not require the robot dynamics. However, they need the Jacobian or the inverse kinematics solution to transform the task-space control input to joint space.

We will discuss a novel method that avoids the inverse kinematics solution and the Jacobian matrix by using minimal representation of the Euler angles. The key idea of the method is to use a linearized version of the Euler angles to decouple the robot control and simplify the task-joint space transformation.

Human-Robot Interaction Control Using Reinforcement Learning, First Edition. Wen Yu and Adolfo Perrusquía.
© 2022 The Institute of Electrical and Electronics Engineers, Inc. Published 2022 by John Wiley & Sons, Inc.

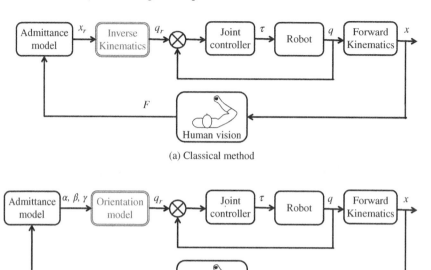

(a) Classical method

(b) Use Euler angles

Figure 5.1 HITL in joint space.

5.2 Joint-Space Control

The human-in-the-loop scheme The HITL scheme in joint space is shown in Figure 5.1. It can be divided into two loops: the inner loop is the joint control, which forces each joint q of the robot to follow the desired angles q_r; and the outer loop is the admittance model, which is used to generate the desired joint angles q_r from the human exerted force/torque f_e.

The outer loop is feedback by human vision, which will serve as reference of the end effector. The desired position x_r and q_r that satisfy the inverse kinematics solution (A.2) and the Jacobian mapping (A.3) is

$$\dot{x}_r = J(q)\dot{q}_r, \quad \dot{q}_r = J^{-1}(q)\dot{x}_r,$$
$$q_r = invk\left(x_r\right). \tag{5.1}$$

The admittance model is

$$\frac{x_r(s)}{f_e(s)} = \frac{1}{M_d s^2 + B_d s + K_d}. \tag{5.2}$$

For this application $x_r = x_d$,

$$M_d \ddot{x}_r + B_d \dot{x}_r + K_d x_r = f_e.$$

Usually, the desired reference has six components, so $x_r(s)$ becomes

$$
\begin{bmatrix} X \\ Y \\ Z \\ \alpha \\ \beta \\ \gamma \end{bmatrix} = \frac{1}{M_d s^2 + B_d s + K_d} \begin{bmatrix} F_x \\ F_y \\ F_z \\ \tau_x \\ \tau_y \\ \tau_z \end{bmatrix},
\tag{5.3}
$$

where X, Y, Z are Cartesian positions; α, β, γ are the Euler angles, which represents the end-effector orientation; F_x, F_y, F_z are the forces; and τ_x, τ_y, τ_z are the torques of the force/torque sensor. Here the forces and torques are decoupled, while the pose and orientation of the robot are coupled. We will use the orientations of the end effector $\alpha, \beta,$ and γ in (5.3).

The orientation of the robot is given by the rotation matrix of the Denavit-Hartenberg (DH) transformation matrix T [11], which is nonlinear. For an HITL control task, the joint angles are

$$
q(t + \Delta t) = q(t) + q_r(t).
\tag{5.4}
$$

When the human is in the control loop, the joint displacements are small at each time step, so q_r is small enough such that $q(t + \Delta t)$ is close to the previous position $q(t)$. Since $q_r(t)$ is small, we can linearize the Euler angles using a Taylor series at the actual robot pose $q(t)$ to avoid robot inverse kinematics $invk\,(\cdot)$,

$$
O = h\,(q(t)) + \sum_{i=1}^{n} \frac{\partial h(q)}{\partial q_i} \left(q_i - q_i(t)\right),
\tag{5.5}
$$

where $O = [\alpha, \beta, \gamma]^{\mathsf{T}}$, $h(q)$ is the Euler angles nonlinear solution. The solutions of $h(q)$ for a non-singular point are:

$$
\alpha = \arctan\left(\frac{r_{32}}{r_{33}}\right), \quad \beta = \arctan\left(\frac{-r_{31}}{\sqrt{r_{11}^2 + r_{21}^2}}\right),
\tag{5.6}
$$

$$
\gamma = \arctan\left(\frac{r_{21}}{r_{11}}\right),
$$

where $r_{i,j}$, $i, j = 1, 2, 3$, are the elements of the rotation matrix of the DH transformation matrix. At a singular point the solutions are

$$
\alpha = \alpha_0, \quad \beta = \pm\frac{\pi}{2}, \quad \gamma = \alpha_0 + \arctan\left(\frac{r_{23}}{r_{22}}\right) \quad \text{if } \sin\beta > 0,
\tag{5.7}
$$

$$
\gamma = -\alpha_0 - \arctan\left(\frac{r_{12}}{r_{13}}\right) \quad \text{if } \sin\beta < 0,
$$

where α_0 is constant orientation (5.6) (see [11]).

(5.7) gives the orientation of the robot at a singular point. We use (5.6) and (5.7) for the linearization of (5.5) to decouple the solutions. (5.7) has the knowledge of the offset α_0, while (5.6) gives the connection between the orientation and the joint angles.

We use the following four types of robots to show how to use the linearized Euler angles for the HITL control.

2-DOF Robot (Pan and Tilt)

The 2-DOF pan and tilt robot is shown in Appendix A.2. The DH transformation matrix T of the pan and tilt robot is

$$
T = \begin{bmatrix}
\cos(q_1)\cos(q_2) & -\cos(q_1)\sin(q_2) & \sin(q_1) & l_2\cos(q_1)\cos(q_2) \\
\sin(q_1)\cos(q_2) & -\sin(q_1)\sin(q_2) & -\cos(q_1) & l_2\sin(q_1)\cos(q_2) \\
\sin(q_2) & \cos(q_2) & 0 & l_1 + l_2\sin(q_2) \\
0 & 0 & 0 & 1
\end{bmatrix},
$$

where l_1 and l_2 are the lengths of the robot link-1 and link-2. The inverse kinematics of the robot is

$$
q_1 = \arctan\left(\frac{Y}{X}\right), \quad q_2 = \arctan\left(\frac{Z - l_1}{\sqrt{l_2^2 - (Z - l_1)^2}}\right),
$$

where X, Y, Z are the robot pose. The above solution requires knowledge of the robot kinematics, and the robot pose is coupled by the joint angle q_2. The forces applied at the force/torque sensor are decoupled. At a non-singularity position, the robot Euler angles in (5.6) are

$$
\begin{aligned}
\alpha &= \tfrac{\pi}{2}, \\
\beta &= -q_2, \\
\gamma &= q_1.
\end{aligned}
\tag{5.8}
$$

When the linearization (5.5) is applied to the Euler angles (5.6) at the actual singular pose $q_0 = \left[0, \tfrac{\pi}{2}\right]^T$, choosing $\alpha_0 = \tfrac{\pi}{2}$ yields the same expression (5.8). This is a one-to-one mapping, i.e., two orientations in (5.8) generate two desired angles in the joint space, so it avoids the inverse kinematics.

4-DOF Robot (Exoskeleton)

The 4-DOF exoskeleton robot is shown in Appendix A.1. The DH transformation matrix T of the exoskeleton robot is

$$
T = \begin{bmatrix}
c_1(c_2c_3c_4 - s_2s_4) - c_4s_1s_3 & s_1s_3s_4 - c_1(c_4s_2 + c_2c_3s_4) & c_3s_1 + c_1c_2s_3 & X \\
c_2c_3c_4s_1 + c_1c_4s_3 - s_1s_2s_4 & -c_4s_1s_2 - (c_2c_3s_1 + c_1s_3)s_4 & -c_1c_3 + c_2s_1s_3 & Y \\
c_3c_4s_2 + c_2s_4 & c_2c_4 - c_3s_2s_4 & s_2s_3 & Z \\
0 & 0 & 0 & 1
\end{bmatrix},
$$

where $s_i = \sin(q_i)$, $c_i = \cos(q_i)$, $i = 1, 2, 3, 4$. The inverse kinematic is obtained by fixing the joint angle $q_3 = 0$,

$$q_1 = \arctan\left(\frac{Y}{X}\right), \quad q_2 = \arctan\left(\frac{\sqrt{1-j^2}}{j}\right) - \arctan\left(\frac{b}{a}\right),$$

$$q_3 = 0, \quad q_4 = \arctan\left(\frac{\sqrt{1-k^2}}{k}\right),$$

where j, k, a, b are functions that depend on kinematic parameters and the robot pose. k, a, and b depend on non-linear functions of the joint q_1 and q_4. The Euler angles in (5.6) at a non-singular point are

$$\alpha = \arctan\left(\frac{c_2c_4-c_3s_2s_4}{s_2s_3}\right),$$

$$\beta = \arctan\left(\frac{-c_3c_4s_2-c_2s_4}{\sqrt{1-\left(c_3c_4s_2+c_2s_4\right)^2}}\right), \tag{5.9}$$

$$\gamma = \arctan\left(\frac{c_2c_3c_4s_1+c_1c_4s_3-s_1s_2s_4}{-c_4s_1s_3+c_1(c_2c_3c_4-s_2s_4)}\right).$$

Here the orientations are strongly coupled. But the linearized angles are decoupled. Since it is a redundant robot, we can fix an extra DOF to certain value. If we fix $q_3 = 0$ in (5.9) and apply the linearization method (5.5) at the actual singular pose $q_0 = \left[0, \frac{\pi}{2}, 0, 0\right]^{\mathsf{T}}$,

$$\alpha = \frac{\pi}{2},$$
$$\beta = -q_2 - q_4, \tag{5.10}$$
$$\gamma = q_1.$$

By a geometrical property we can see that at the actual pose q_3 affects the α orientation. The linearized orientation equations of the 4-DOF robot are

$$\alpha = \frac{\pi}{2} - q_3,$$
$$\beta = -q_2 - q_4, \tag{5.11}$$
$$\gamma = q_1.$$

The advantage of using this linearization is that the orientations are decoupled, and they have direct relations with the joint angles when DOFs is equal or less than three. When the DOFs are more than three, we can use one of the following methods to generate the desired joint angles q_r from the orientations.

1. The three orientations (α, β, γ) are obtained from the torques (τ_x, τ_y, τ_z) of the force/torque sensor. The first three joint angles are from these three orientations, and the other joint angles can be estimated from the forces components (F_x, F_y, F_z) of the force/torque sensor, such that the linearization (5.5) is satisfied.

2. The operator first moves the robot to the desired position with the first three joint angles, and then the operator moves the robot to the desired orientation angles by switching the orientation components. In fact, this two-step method is applied in many industrial robots.

But there exists a joint linear combination at the orientation β in (5.11). We first divide the orientation β into two parts: $\beta = \beta_{q_2} + \beta_{q_4}$. Then we use both torque and force at the y-direction to generate β_{q_2} and β_{q_4}. The desired joint positions from the force/torque sensor are

$$
\begin{aligned}
q_3 &= \tfrac{\pi}{2} - \alpha, & \alpha &= admit\left(\tau_x\right), \\
q_1 &= \gamma, & \gamma &= admit\left(\tau_z\right), \\
q_2 &= -\beta_{q_2}, & \beta_{q_2} &= admit\left(\tau_y\right), \\
q_4 &= -\beta_{q_4}, & \beta_{q_4} &= admit\left(F_y\right),
\end{aligned}
\tag{5.12}
$$

where $admit\left(\cdot\right)$ is the admittance model (5.3).

5-DOF Robot (Exoskeleton)

The orientations of the 5-DOF exoskeleton robot end effector is obtained in a similar way as for the 4-DOF robot. The relations of the orientations with the joint angles are

$$
\begin{aligned}
\alpha &= \tfrac{\pi}{2} - q_3, \\
\beta &= -q_2 - q_4, \\
\gamma &= q_1 + q_5,
\end{aligned}
\tag{5.13}
$$

where β and γ are linear combinations of the joint angles. We can use a similar method as the 4-DOF robot to separate the orientations into two terms and use an additional force components as

$$
\begin{aligned}
q_3 &= \tfrac{\pi}{2} - \alpha, & \alpha &= admit\left(\tau_x\right), \\
q_2 &= \beta_{q_2}, & \beta_{q_2} &= admit\left(\tau_y\right), \\
q_1 &= \gamma_{q_1}, & \beta_{q_1} &= admit\left(\tau_z\right), \\
q_4 &= -\beta_{q_4}, & \beta_{q_4} &= admit\left(F_y\right), \\
q_5 &= \gamma_{q_5}, & \gamma_{q_5} &= admit\left(Fz\right).
\end{aligned}
\tag{5.14}
$$

6-DOF Robot (Exoskeleton)

If the second joint angle of the 5-DOF exoskeleton robot is a spherical wrist, it becomes a 6-DOF robot. The orientations satisfy the following relations:

$$
\begin{aligned}
\alpha &= \tfrac{\pi}{2} - q_3 - q_6, \\
\beta &= -q_2 - q_4, \\
\gamma &= q_1 + q_5.
\end{aligned}
\tag{5.15}
$$

Here all orientations are linear combinations of the joint angles, and each of them is divided into two terms. Solving for the joint angles yields

$$
\begin{aligned}
q_3 &= \tfrac{\pi}{2} - \alpha_{q_3}, & \alpha_{q_3} &= admit\left(\tau_x\right) \\
q_6 &= -\alpha_{q_6}, & \alpha_{q_6} &= admit\left(F_x\right) \\
q_2 &= -\beta_{q_2}, & \beta_{q_2} &= admit\left(\tau_y\right) \\
q_4 &= -\beta_{q_4}, & \alpha_{q_4} &= admit\left(F_y\right) \\
q_1 &= \gamma_{q_1}, & \gamma_{q_1} &= admit\left(\tau_z\right) \\
q_5 &= \gamma_{q_5}, & \gamma_{q_5} &= admit\left(F_z\right)
\end{aligned}
\tag{5.16}
$$

In view of the above examples, we can write (5.5) as a linear combination of the joint angles as

$$
\alpha = \sum_{i=1}^{n}(c_i q_i + d_i), \quad \beta = \sum_{i=1}^{n}(c_i q_i + d_i), \quad \gamma = \sum_{i=1}^{n}(c_i q_i + d_i),
\tag{5.17}
$$

where c_i and d_i are offsets of the orientation, which depend of the robot configuration, q_i is the i-th joint angle, and $q = \begin{bmatrix} q_1 & \cdots & q_n \end{bmatrix}^{\mathsf{T}}$. There exist several ways to write (5.17). However, for HITL control, it is not relevant, because we have small displacements in each step and the human moves the robot into the desired position through the admittance model.

5.3 Task-Space Control

The task-space schemes are shown in Figure 5.2. The Jacobian matrix provides the mapping between the joint space and the task space, which can be expressed as

$$
J(q) = \left[J_v^{\mathsf{T}}(q), J_\omega^{\mathsf{T}}(q) \right]^{\mathsf{T}},
\tag{5.18}
$$

where $J_v(q)$ is the linear velocity Jacobian, and $J_\omega(q)$ is the angular velocity Jacobian. $J_v(q)$ requires kinematic parameters of the robot and joint measurements, while $J_\omega(q)$ only requires joint measurements. We use two types of Jacobian: analytic Jacobian $J_a = J$ and the geometric Jacobian $J_g = J(q)$ in (5.18) [11]. The analytic Jacobian J_a is the differential version from forward kinematics. The relation between the geometric and analytic Jacobians (A.5) is

$$
J_a = \begin{bmatrix} I & 0 \\ 0 & R(O)^{-1} \end{bmatrix} J(q),
\tag{5.19}
$$

where $R(O)$ is a rotation matrix of the orientation components, $O = [\alpha, \beta, \gamma]^{\mathsf{T}}$. For the end effector, the linear velocity component of the analytic Jacobian is the same

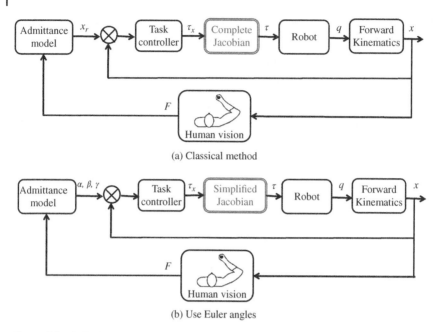

(a) Classical method

(b) Use Euler angles

Figure 5.2 HITL in task space.

as the geometric Jacobian. For the Euler angles approach, we will use the angular velocity component of the analytic Jacobian J_a. Taking the time derivative of (5.17) yields

$$\dot{\alpha} = \sum_{i=1}^{n} c_i \dot{q}_i, \quad \dot{\beta} = \sum_{i=1}^{n} c_i \dot{q}_i, \quad \dot{\gamma} = \sum_{i=1}^{n} c_i \dot{q}_i. \tag{5.20}$$

Note that the relation between the time derivative of the Euler angles and the joint velocities only lies in c_i. The angular velocity component of the analytic Jacobian can be expressed as

$$\left[\dot{\alpha}, \dot{\beta}, \dot{\gamma} \right]^{\mathsf{T}} = J_{a\omega} \dot{q}, \tag{5.21}$$

where $J_{a\omega}$ is the angular velocity component of the analytic Jacobian J_a, which is a constant matrix with c_i in its components. We will use $J_{a\omega}^{\mathsf{T}}$ to calculate the control torque as in (3.39).

We use four types of robot configurations to show the use of the time derivative of Euler angles in task space HITL.

2-DOF Robot (Pan and Tilt)

The 2-DOF pan and tilt robot (see Appendix A.2) has the following angular velocity Jacobian

$$J_\omega(q) = \begin{bmatrix} 0 & \sin(q_1) \\ 0 & -\cos(q_1) \\ 1 & 0 \end{bmatrix}. \tag{5.22}$$

This Jacobian is easy to compute and requires only joint measures. We take the time derivative of (5.8) to obtain the angular velocity of the analytic Jacobian (5.21):

$$\begin{bmatrix} \dot{\alpha} \\ \dot{\beta} \\ \dot{\gamma} \end{bmatrix} = \begin{bmatrix} 0 \\ -\dot{q}_2 \\ \dot{q}_1 \end{bmatrix}.$$

So the angular velocity of the analytic Jacobian is

$$J_{a\omega} = \begin{bmatrix} 0 & 0 \\ 0 & -1 \\ 1 & 0 \end{bmatrix}. \tag{5.23}$$

Notice the the analytic Jacobian in (5.23) has a similar structure to the geometric Jacobian in (5.22), and it is equivalent to evaluating the Jacobian at the actual pose.

4-DOF Robot (Exoskeleton)

The 4-DOF exoskeleton robot (see Appendix A.1) has the following angular velocity Jacobian

$$J_\omega(q) = \begin{bmatrix} 0 & s_1 & -c_1 s_2 & c_1 c_2 s_3 + c_3 s_1 \\ 0 & -c_1 & -s_1 s_2 & c_2 s_1 s_3 - c_1 c_3 \\ 1 & 0 & c_2 & s_2 s_3 \end{bmatrix}. \tag{5.24}$$

Using the same method as the 2-DOF robot,

$$\dot{\alpha} = -\dot{q}_3, \quad \dot{\beta} = -\dot{q}_2 - \dot{q}_4, \quad \dot{\gamma} = \dot{q}_1.$$

So

$$J_{a\omega} = \begin{bmatrix} 0 & 0 & -1 & 0 \\ 0 & -1 & 0 & -1 \\ 1 & 0 & 0 & 0 \end{bmatrix}. \tag{5.25}$$

5-DOF Robot

Taking the time derivative of (5.13), the angular velocity component of the analytic Jacobian is

$$
J_{a\omega} = \begin{bmatrix} 0 & 0 & -1 & 0 & 0 \\ 0 & -1 & 0 & -1 & 0 \\ 1 & 0 & 0 & 0 & 1 \end{bmatrix}. \tag{5.26}
$$

6-DOF Robot

Taking the time derivative of (5.16), the angular velocity component of the analytic Jacobian is

$$
J_{a\omega} = \begin{bmatrix} 0 & 0 & -1 & 0 & 0 & -1 \\ 0 & -1 & 0 & -1 & 0 & 0 \\ 1 & 0 & 0 & 0 & 1 & 0 \end{bmatrix}. \tag{5.27}
$$

In order to avoid the pose coupling, we only use the orientation of the robot. However, when the orientations are linear combinations of more than two joint angles, there are multiple solutions of the joint angles. To solve this problem we can use one of the following methods:

1) Modify the parameters c_i such that all joint movements contribute to the orientations. Let's consider the 4-DOF robot; the angular velocity component of the analytic Jacobian is changed as follows:

$$
\hat{J}_{a\omega} = \begin{bmatrix} 0 & 0 & -1 & 0 \\ 0 & -0.8 & 0 & -0.2 \\ 1 & 0 & 0 & 0 \end{bmatrix},
$$

where $\hat{J}_{a\omega}$ is an approximation of $J_{a\omega}$. This estimation is reliable if the human (operator) knows the movement contribution of each joint angle, e.g., the first three DOF determine the robot position, which requires a big contribution, while the last 3 DOF determine the orientation, which requires less contribution at the Jacobian matrix.

2) As in joint space, we can divide the Jacobian in two parts: $\hat{J}_{a\omega} = \hat{J}_{a\omega_1} + \hat{J}_{a\omega_2}$. For the 4-DOF robot,

$$
\hat{J}_{a\omega_1} = \begin{bmatrix} 0 & 0 & -1 & 0 \\ 0 & -1 & 0 & 0 \\ 1 & 0 & 0 & 0 \end{bmatrix}, \quad \hat{J}_{a\omega_2} = \begin{bmatrix} 0 & 0 & 0 & 0 \\ 0 & 0 & 0 & -1 \\ 0 & 0 & 0 & 0 \end{bmatrix}. \tag{5.28}
$$

The first Jacobian gives the robot position, and the second Jacobian gives a desired orientation. Since the orientation model has complete relation with the joint angles and these are decoupled, the singularities are avoided.

5.4 Experiments

In order to test the Euler angles approach in both joint space and task space, we use two robots: the 2-DOF pan and tilt robot shown in Figure 5.3 and the 4-DOF exoskeleton shown in Figure 5.4. Both robots are controlled by the human operator with the Schunk force/torque (F/T) sensor. The real-time environment is the Simulink and MATLAB 2012. The communication protocol is the controller area network (CAN bus), which enables the PC to communicate with the actuators and the F/T sensor. For both joint and task space, the controller gains are tuned manually until satisfactory responses are obtained.

The human applies certain amount of force/torque at the sensor to generate random movements using the admittance model. We compare the classical admittance control in joint space (3.23) and task space (3.3) with the model-free

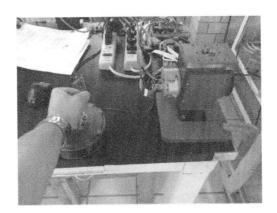

Figure 5.3 2-DOF pan and tilt robot.

Figure 5.4 4-DOF exoskeleton robot.

Table 5.1 Controller gains for the pan and tilt robot

Gain	PD joint\task	Adaptive PD joint\task	PID joint\task	Sliding PD joint\task
K_p	$5 \times 10^3 \backslash 5 \times 10^4$	-	-	-
K_v	100	-	-	-
Ψ	-	-	0.15	0.15\-
Λ	-	90	90	90
K_s	-	0.9	0.9	0.9
K_Θ	-	1×10^{-3}	-	-
k_1	-	-	-	-\0.2
k_2	-	-	-	-\0.5

Table 5.2 Controller gains for the exoskeleton

Gain	PD joint\task	Adaptive PD joint\task	PID joint\task	Sliding PD joint\task
K_p	$2 \times 10^3 \backslash 4 \times 10^4$	-	-	-
K_v	80	-	-	-
Ψ	-	-	0.3	0.5\-
Λ	-	180	200	90
K_s	-	01.2	0.9	1.2
K_Θ	-	7×10^{-3}	-	-
k_1	-	-	-	-\1.6
k_2	-	-	-	-\1.3

controllers in joint space: adaptive PD (4.58), PID (4.60), and sliding PD (4.61); and the task space controllers: adaptive PD (4.54), PID (4.55), and sliding PD (4.56).

The dynamics of the robot is partially unknown, and the robots have high friction in each joint. The control gains of the pan and tilt robot and the exoskeleton are given in Table 5.1 and Table 5.2, respectively.

The following admittance model for the torques components of the force/torque sensor is used, such that it satisfies (5.4),

$$admit(\cdot) = \frac{1}{M_{d_i} s^2 + B_{d_i} s + K_{d_i}},$$

where $M_{d_i} = 1$, $B_{d_i} = 140$, and $K_{d_i} = 4000$, with $i = 1, 2, 3$.

The force components are scaled with a factor of 1:30 so that they are within the same range of the torques components. The admittance model is the same for both robots since its parameters depend on the sensor, not on the robot configuration. The controller gains are tuned manually for both joint space and task space and are given in Table 5.1. The classical admittance controllers (PD joint and PD task) have big gains due the modelling error.

Example 5.1 *2-DOF Pan and Tilt Robot*

We test the joint and task space controllers of the 2-DOF pan and tilt robot. The human applies torques to move the robot to desired positions. These torques are the same for both joint and task space controllers. The inverse kinematics is avoided by (5.8). Figure 5.5 shows the experimental results of the joint space controllers. Since they do not use the inverse kinematics, there are not singularity problems. The classical admittance control has accuracy error due the modelling error, but the model-free controllers are robust. Here the orientations are completely decoupled and do not require a force component. The task space controllers of the pan and tilt robot use the analytic angular velocity Jacobian (5.23) for the control torque. Figure 5.6 shows the experiment results of the task space control.

Since the Jacobian only has only 0s and ±1s in its components, its contribution only lies in the control torque transformation, and for this robot in particular, the control solution is practically the same as in joint space. This also avoids singularity problems. The model-free controllers achieve the reference tracking without knowledge of the robot dynamics.

Example 5.2 *4-DOF Exoskeleton Robot*

Now consider the 4-DOF exoskeleton robot. The inverse kinematics is avoided by (5.11). This robot is redundant, and the inverse kinematics requires to fix one joint angle to calculate the other joint angles. (5.11) avoids this problem by using the orientation components of the end effector. The comparison results are given in Figure 5.7 and Figure 5.8.

Here the orientation is a linear combination of the joint angles, and the force and torque components are used to decouple the orientation into two terms as in (5.12). Also the orientation linear model avoids singularity problems and knowledge of the kinematic parameters. Since the robot has high gravity terms, the classical admittance control has a big error in joint positions q_2 and q_4. The model-free controllers are robust and overcome this problem.

For the task space controllers, the analytic angular velocity Jacobian (5.25) has a linear combination of two joint angles. To solve this problem, we use the Jacobian approximation (5.28). The key idea is to move the robot to a desired pose and then change the Jacobian of the end effector to a desired orientation. The comparison results are given in Figure 5.9.

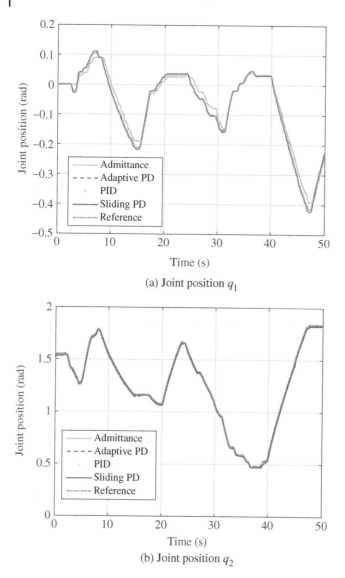

(a) Joint position q_1

(b) Joint position q_2

Figure 5.5 Control of pan and tilt robot in joint space.

The model-free controllers present better results than classical admittance control. For the HRI scheme, the inverse kinematics and the exact Jacobian matrix are not needed by means of human vision. The singularity problem and instability are avoided because each joint angle and orientation are decoupled.

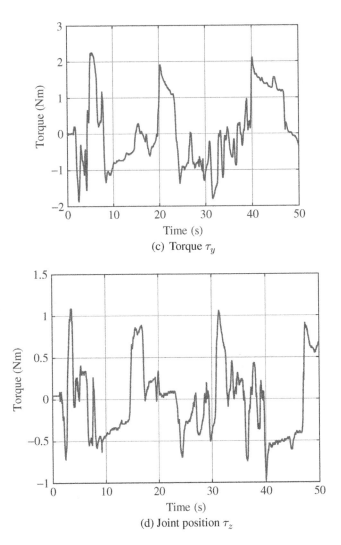

(c) Torque τ_y

(d) Joint position τ_z

Figure 5.5 (*Continued*)

Discussion

In joint space, when the dynamics of the robot are known and there are no disturbances, the classical admittance control can guarantee the position tracking. When
the dynamics of the robot are unknown and there are disturbances, the classical
admittance control becomes poor.

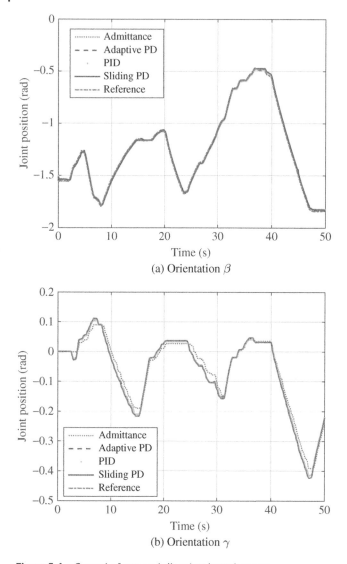

(a) Orientation β

(b) Orientation γ

Figure 5.6 Control of pan and tilt robot in task space.

The adaptive and sliding mode PD controllers have good performances for the control task. The adaptive PD control shows good and smooth responses; however, it requires model structure. The sliding mode PD is a model-free controller that shows good response and is robust, but it presents the chattering problem,

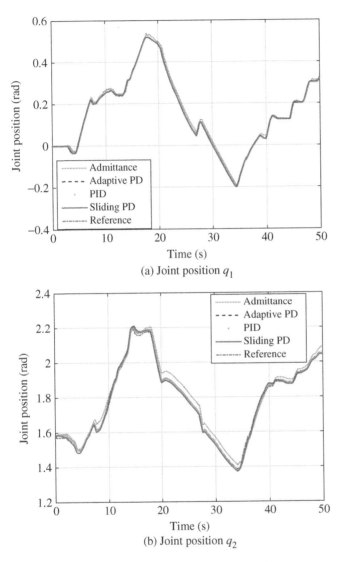

Figure 5.7 Control of 4-DOF exoskeleton robot in joint space.

which is not reliable for human-robot cooperation tasks. The second-order sliding mode controllers (4.56) and (4.62) overcome this problem. The PID control is also a model-free controller whose performance is good, but its integral gain must be adjusted carefully to avoid bad transient performance.

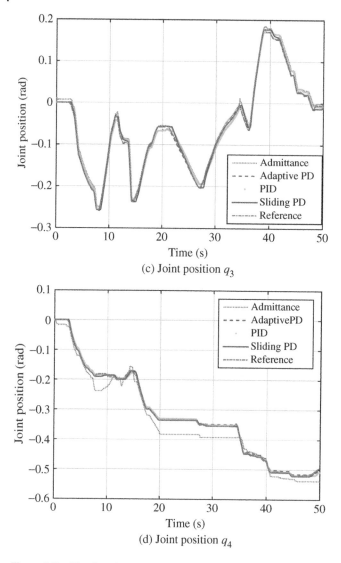

(c) Joint position q_3

(d) Joint position q_4

Figure 5.7 *(Continued)*

Limitations and Scope

These approaches are reliable for cooperative tasks between humans and robots. Condition (5.4) simplifies the orientation model. When the robot has 7 or more DOFs, the Euler angles approach can not be applied directly, because the redundant robot has more DOFs than the Cartesian degrees.

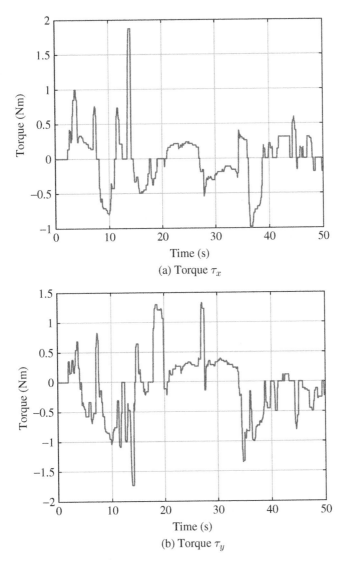

Figure 5.8 Torques and forces of 4-DOF exoskeleton robot.

When the DOF is equal to or less than 6, the Euler angles approach can facilitate the human-robot cooperation task [6, 12]. When the human is in contact with the robot, careful design of the admittance model and the use of inverse kinematics or the Jacobian matrix are required [13, 14].

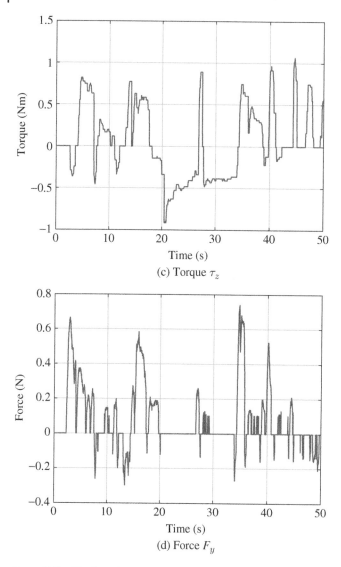

(c) Torque τ_z

(d) Force F_y

Figure 5.8 (Continued)

5.5 Conclusions

In this chapter the HITL robot interaction control is studied. The main problem of this application is the use of the inverse kinematics and the Jacobian matrix, which are not always available and may have singularities problems. To solve this

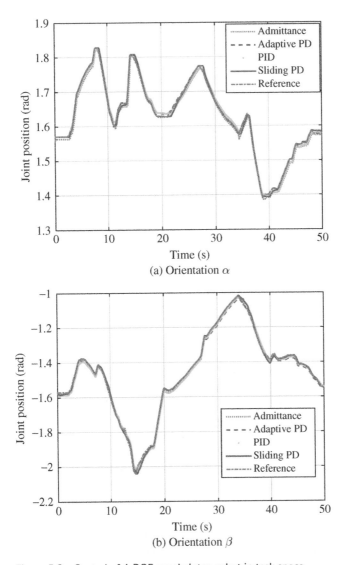

Figure 5.9 Control of 4-DOF exoskeleton robot in task space.

issue, the approach using Euler angles is given. The Euler angles approach defines relations between the orientation components and the joint angles, and it does not need the inverse kinematics or the complete Jacobian matrix of the robot. Also the classical and the model-free admittance controllers are modified with the Euler angles approach. Experiments use the 2-DOF pan and tilt robot and the 4-DOF exoskeleton. The results show the effectiveness of the Euler angles approach.

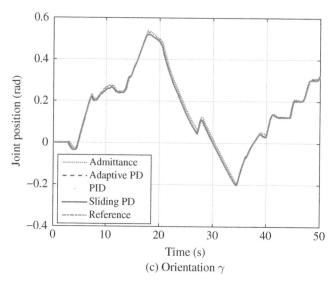

(c) Orientation γ

Figure 5.9 (*Continued*)

References

1 F. Ficuciello, L. Villani, and B. Siciliano, "Variable Impedance Control of Redundant Manipulators for Intuitive Human-Robot Physical Interaction," *IEEE Transactions on Robotics*, vol. 31, pp. 850–863, August 2015.

2 R. Bonitz and T. Hsia, "Internal Force-Based Impedance Control for Cooperating Manipulators," *IEEE Transactions on Robotics and Automation*, vol. 12, pp. 78–89, February 1996.

3 A. Abdossalami and S. Sirouspour, "Adaptive Control of Haptic Interaction with Impedance and Admittance Type Virtual Environments," *IEEE*, pp. 145–152, March 13–14 2008. Symposium on Haptic Interfaces for Virtual Environments and Teleoperator Systems.

4 H. Kazerooni and M. Her, "The Dynamics and Control of a Haptic Interface Device," *IEEE Transactions on Robotics and Automation*, vol. 10, pp. 453–464, August 1994.

5 J. Garrido, *Aprendizaje por demostración en el espacio articular para el seguimiento de trayectorias aplicado en un exoesqueleto de 4 grados de libertad.* PhD thesis, Centro de Investigación y Estudios Avanzados del Instituto Politécnico Nacional, México DF, enero 2015.

6 M. Dimeas and N. Aspragathos, "Online stability in human-robot cooperations with admittance control," *IEEE Transactions on Haptics*, vol. 9, no. 2, pp. 267–278, 2016.

7 W. Yu, J. Rosen, and X. Li, "PID Admittance Control for an Upper Limb Exoskeleton," *American Control Conference*, pp. 1124–1129, June 29–July 01 2011. O'Farrel Street, San Francisco, CA, USA.

8 N. Hogan, "Impedance Control: An Approach to Manipulation," *Journal of Dynamic Systems, Measurement, and Control*, vol. 107, pp. 1–24, March 1985. Transactions of the ASME.

9 W. Yu, R. Carmona Rodriguez, and X. Li, "Neural PID Admittance Control of a Robot," *American Control Conference*, pp. 4963–4968, June 17–19 2013. Washington, DC, USA.

10 S. O. Onyango, "Behaviour modelling and system control with human in the loop," *Diss. Université Paris-Est*, 2017.

11 M. W. Spong, S. Hutchinson, and M. Vidyasagar, *Robot Dynamics and Control*. John Wiley & Sons, Inc., Second ed., January 28, 2004.

12 W. Yu and J. Rosen, "A Novel Linear PID Controller for an Upper Limb Exoskeleton," *49th IEEE Conference on Decision and Control*, pp. 3548–3553, December 2010. Atlanta, Hotel, GA, USA.

13 S. Roy and Y. Edan,"Investigating joint-action in short-cycle repetitive handover tasks: The role of giver versus receiver and its implications for human-robot collaborative system design," *International Journal of Social Robotics*, pp. 1–16, 2018.

14 R. Someshwar and Y. Kerner, "Optimization of waiting time in hr coordination," *2013 IEEE International Conference on Systems, Man, and Cybernetics (SMC13)*, pp. 1918–1923, 2013.

Part II

Reinforcement Learning for Robot Interaction Control

6

Reinforcement Learning for Robot Position/Force Control

6.1 Introduction

As we have seen in the previous chapters, the most popular approaches for interaction control [1, 2] are impedance or admittance control and position/force control. In this chapter we will focus on the position/force control problem. The position/force control has two control loops:

1) The internal loop, which is the position control.
2) The external loop, which is the force control.

The force control is designed as an impedance control [3]. The performance of the impedance control depends on the impedance parameters [4], which come from the dynamics of the environment [5].

The input of the impedance model is the position/velocity of the end effector, and its output is a force. Therefore, it prevents a certain amount of force from moving in the environment [6]. The position/force control uses the impedance model to generate the desired force [7]. The objective of the position/force is that the output force is as small as possible and that the position is as close as its reference. This problem can be solved with the linear quadratic regulator (LQR) [8, 9]. It gives the optimal contact force and minimizes the position error. However, LQR needs the dynamic model of the environment.

When the environment is unknown, impedance control does not work well for interaction performance [10–12], and the force control is not robust to generate the desired force with respect to uncertainties in the environment [5, 13]. There are several methods to estimate the dynamics of the environment as we discussed

Human-Robot Interaction Control Using Reinforcement Learning, First Edition. Wen Yu and Adolfo Perrusquía.
© 2022 The Institute of Electrical and Electronics Engineers, Inc. Published 2022 by John Wiley & Sons, Inc.

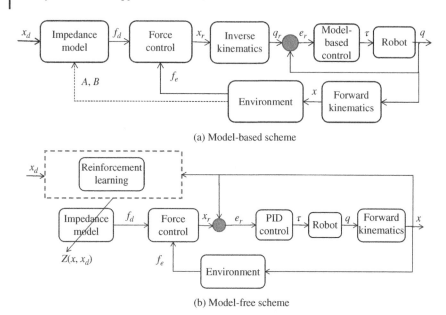

(a) Model-based scheme

(b) Model-free scheme

Figure 6.1 Position/force control.

in Chapter 2. To obtain the position/force control, there are three steps (see Figure 6.1(a)):

1) Estimation of environmental parameters;
2) The design of the desired impedance model via the LQR controller;
3) Application of the position/force control [14].

One way to avoid the above steps is to use reinforcement learning, which will be discussed in the following sections.

6.2 Position/Force Control Using an Impedance Model

Assume that we have knowledge of the environment parameters. The model-based robot interaction control scheme is shown in Figure 6.1(a). The outer loop is the force control, which has three blocks: the environment, the impedance model, and the force controller. The impedance model is designed according to the environment parameters (A, B) and the desired position reference x_d. The LQR controller gives the optimal desired force f_d^*, which generates the new position reference x_r. The inner loop is the position control, which has three blocks: the inverse kinematics, the model-based control, and the robot. The inverse kinematics of the robot is

Figure 6.2 Robot-environment
interaction

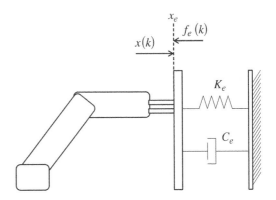

used to transform the output of the force controller to a joint-space reference q_r.
Then a model-based controller is used to guarantee position tracking.

The robot dynamics (A.17) is continuous. The discretization uses the integration
method. We define k as the rate of time step, and let T represent the sampling time
interval. Then the joint angles are $q(k) = q(t_k)$, at time $t_k = kT$.

When the robot end effector touches the environment, the force f_e is generated
(see Figure 6.2). The environment can be modeled as (2.24):

$$f_e = -C_e \dot{x} + K_e \left(x_e - x \right),$$

(6.1)

where $C_e, K_e \in \mathbb{R}^{m \times m}$ are the environment damping and stiffness constant matrices, respectively, and $x_e \in \mathbb{R}^m$ is the position of the environment.

Let $m = 1$. When $x < x_e$, there is no interaction force, on the other hand. Here
$x \geq x_e$, and there is an interaction force between the robot end effector and the
environment. Model (6.1) can be written in discrete time as follows:

$$x_{k+1} = A_e x_k + B_e f_{e_k}, \ x_k \geq x_e,$$

(6.2)

where $A_e = -C_e^{-1} \left(K_e - C_e \right)$, and $B_e = -C_e^{-1}$. The typical target impedance model
is a function of the error between the desired position and the virtual reference
[6]:

$$Z(e_{r_k}) = f_{d_k},$$
$$Z(x_{r_k}, x_{d_k}) = f_{d_k},$$

(6.3)

where $Z(\cdot)$ is the impedance model, which is designed according to the environ-
ment model, $e_{r_k} = x_{r_k} - x_d$ is the position error, $x_{r_k} \in \mathbb{R}^m$ is the virtual reference
trajectory, and $x_d \in \mathbb{R}^m$ is the desired constant reference. Both references are in
Cartesian space.

The main objective is to determine the optimal desired force $f_{d_k}^*$, which is the
minimum force that minimize the position error e_{r_k}. Note that the environment

model is a first-order system whose control is f_{e_k}, so the impedance model has the same structure as (6.2),

$$e_{r_{k+1}} = A_e e_{r_k} + B_e f_{d_k}.$$ (6.4)

The classical PID control in task space is given by

$$\tau = J^T(q) f_\tau,$$ (6.5)

$$f_\tau = K_p e + K_v \dot{e} + K_i \int_0^t e(\sigma) d\sigma + f_e,$$ (6.6)

where $e = x_r - x$ is the tracking error, and $K_p, K_v, K_i \in R^{m \times m}$ are the proportional, derivative, and integral gains of the PID controller, respectively. Notice that this controller is different to the PID control law (4.60). In discrete time,

$$f_{\tau_k} = K_p e_k + K_v \left[e_{k+1} - e_k \right] + K_i \sum_{i=1}^k e_i + f_{e_k}.$$ (6.7)

Since the PID controller is designed such that it satisfies

$$\lim_{k \to \infty} x_k = x_{r_k},$$

the position error is rewritten as

$$e_{r_k} = x_k - x_d.$$

Since the force measurement is available, the proposed force control is an admittance control as follows:

$$x_r(s) = (sI + B_a^{-1} K_a)^{-1} \left(x_r(0) + x_d(s) + B_a^{-1} e_f(s) \right),$$ (6.8)

where $e_f(s) = f_e(s) - f_d(s)$ is the force error between the desired force f_d and the external force component f_e, and $B_a, K_a \in R^{m \times m}$ are the damping and stiffness of the admittance model.

The goal of the impedance model is to generate an optimal force $f_{d_k}^*$ when the robot interacts with the environment that minimizes the position error. We want to design an optimal control as

$$f_{d_k}^* = L e_{r_k},$$ (6.9)

where $L \in R^{m \times m}$ is the gain that minimizes the discounted cost function

$$J(k) = \sum_{i=k}^{\infty} \gamma^{i-k} \left(e_{r_i}^T S e_{r_i} + f_{d_i}^T R f_{d_i} \right),$$ (6.10)

where $\gamma \in (0, 1]$ is a discount factor, $S \in \mathbb{R}^{m \times m}$ and $R \in \mathbb{R}^{m \times m}$ are weights matrices, $S = S^T \geq 0$, and $R = R^T > 0$. S is the weight of the state, and R is the weight of control input.

The above cost function is the tracking problem, which corresponds to the general case of the optimal control problem. The gain vector L can be calculated by using the discrete-time algebraic Riccati equation (DARE) [15]:

$$A_e^\mathsf{T} P A_e - P + S - A_e^\mathsf{T} P B_e \left(B_e^\mathsf{T} P B_e + \frac{1}{\gamma} R \right)^{-1} B_e^\mathsf{T} P A_e = 0, \tag{6.11}$$

where P is the solution of the DARE. Then the feedback control gain L is given by

$$L = - \left(B_e^\mathsf{T} P B_e + \frac{1}{\gamma} R \right)^{-1} B_e^\mathsf{T} P A_e. \tag{6.12}$$

Since the environment model is a first-order system, the controller only requires the virtual stiffness gain [16].

6.3 Reinforcement Learning Based Position/Force Control

If we do not have environment parameters, we use a model-free control scheme, which is shown in Figure 6.1(b). The outer loop is the force control, which uses the optimal impedance model $Z^*(x, x_d)$ and gives the optimal desired force f_d^* together with the exerted force f_e. The force controller generates the new reference position x_r. The inner loop is the position control, which uses a PID controller in task space [17] to avoid the inverse kinematics and knowledge of the robot dynamics. Unlike the classical model-based controller, as in Figure 6.1(a), the model-free controller does not require additional information to compute the optimal desired force.

We do not need to estimate the environment parameters for the design of the impedance model; we use the external force f_e and the admittance control. It is an indirect way of knowing the dynamic properties of the environment. The impedance model requires knowledge of the dynamics of the environment to obtain the optimal solution. If we do not have previous information on the dynamics of the environment, we use the reinforcement learning to obtain the desired force. We can define the following discounted cost-to-go value function $V(e_{r_k})$:

$$V(e_{r_k}) = \sum_{i=k}^{\infty} \gamma^{i-k} \left(e_{r_i}^\mathsf{T} S e_{r_i} + f_{d_i}^\mathsf{T} R f_{d_i} \right)$$

$$= e_{r_k}^\mathsf{T} S e_{r_k} + f_{d_k}^\mathsf{T} R f_{d_k} + \sum_{i=k+1}^{\infty} \gamma^{i-k} \left(e_{r_i}^\mathsf{T} S e_{r_i} + f_{d_i}^\mathsf{T} R f_{d_i} \right)$$

$$= r_{k+1} + \gamma V(e_{r_{k+1}}), \tag{6.13}$$

where

$$r_{k+1} = e_{r_k}^T S e_{r_k} + f_{d_k}^T R f_{d_k},$$

where r_{k+1} is called the reward function or utility function. The value function V is minimized by finding the optimal control policy (or optimal desired force),

$$f_{d_k}^* = \arg \min_{f_{d_k}} V\left(e_{r_k}\right). \tag{6.14}$$

Assuming that the optimal control exists, the optimal value function satisfies

$$
\begin{aligned}
V^*\left(e_{r_k}\right) &= \min_{f_{d_k}} V\left(e_{r_k}\right) \\
&= e_{r_k}^T S e_{r_k} + f_{d_k}^{*T} R f_{d_k}^* + \gamma V^*(e_{r_{k+1}}) \\
&= e_{r_k}^T P e_{r_k},
\end{aligned} \tag{6.15}
$$

where P is the solution of the DARE. The cost-to-go value function can be defined as

$$
\begin{aligned}
V\left(e_{r_k}\right) &= r_{k+1} + \gamma V^*(e_{r_{k+1}}) \\
&= e_{r_k}^T S e_{r_k} + f_{d_k}^T R f_{d_k} + e_{r_{k+1}}^T \gamma P e_{r_{k+1}} \\
&= \begin{bmatrix} e_{r_k} \\ f_{d_k} \end{bmatrix}^T \left\{ \begin{bmatrix} S & 0 \\ 0 & R \end{bmatrix} + \begin{bmatrix} A_e^T \\ B_e^T \end{bmatrix} \gamma P \begin{bmatrix} A_e^T \\ B_e^T \end{bmatrix}^T \right\} \begin{bmatrix} e_{r_k} \\ f_{d_k} \end{bmatrix} \\
&= \begin{bmatrix} e_{r_k} \\ f_{d_k} \end{bmatrix}^T H \begin{bmatrix} e_{r_k} \\ f_{d_k} \end{bmatrix},
\end{aligned} \tag{6.16}
$$

where $H \geq 0$ is a positive semi-definite parameter matrix:

$$H = \begin{bmatrix} \gamma A_e^T P A_e + S & \gamma A_e^T P B_e \\ \gamma B_e^T P A_e & \gamma B_e^T P B_e + R \end{bmatrix} = \begin{bmatrix} H_{11} & H_{12} \\ H_{21} & H_{22} \end{bmatrix}. \tag{6.17}$$

Applying $\partial V(e_{r_k})/\partial f_{d_k} = 0$, the optimal desired control is

$$
\begin{aligned}
f_{d_k}^* &= -H_{22}^{-1} H_{21} e_{r_k} \\
&= -\left(B_e^T P B_e + \frac{1}{\gamma} R \right)^{-1} B_e^T P A_e e_{r_k} = L e_{r_k}.
\end{aligned} \tag{6.18}
$$

Substituting the optimal control (6.18) into (6.16), we find the relationship between H and P:

$$P = \begin{bmatrix} I & L^T \end{bmatrix} H \begin{bmatrix} I & L^T \end{bmatrix}^T. \tag{6.19}$$

We define the cost-action value function Q as

$$Q(e_{r_k}, f_{d_k}) = V(e_{r_k}) = \begin{bmatrix} e_{r_k} \\ f_{d_k} \end{bmatrix}^T H \begin{bmatrix} e_{r_k} \\ f_{d_k} \end{bmatrix}. \tag{6.20}$$

This is the Q-function. The Bellman optimal equation with Q-function is

$$V^*(e_{r_k}) = \min_{f_{d_k}} Q(e_{r_k}, f_{d_k}), \tag{6.21}$$

$$f_{d_k}^* = \arg\min_{f_{d_k}} Q(e_{r_k}, f_{d_k}). \tag{6.22}$$

Then the optimal control policy satisfies the following temporal difference (TD) equation:

$$Q^*(e_{r_k}, f_{d_k}^*) = r_{k+1} + \gamma Q^*(e_{r_{k+1}}, f_{d_{k+1}}^*). \tag{6.23}$$

The optimization problem (6.21) is in fact a Markov decision process (MDP). It can be solved by dynamic programming, which depends on dynamic knowledge. If we do not have dynamic information, the Q-value in (6.23) is not available. We will use approximation methods, such as Monte Carlo, Q-learning, and Sarsa learning, to obtain the approximation of the Q-value.

Giving the control policy

$$h(e_{r_k}) = Le_{r_k} = f_{d_k},$$

its value function is

$$V^h(e_{r_k}) = Q^h(e_{r_k}, f_{d_k}) = \sum_{i=k}^{\infty} \gamma^{i-k} r_{k+1}. \tag{6.24}$$

The recursive form of (6.13) is

$$V^h(e_{r_k}) = r_{k+1} + \gamma V^h(e_{r_{k+1}}), \tag{6.25}$$

$$Q^h(e_{r_k}, f_{d_k}) = r_{k+1} + \gamma Q^h(e_{r_{k+1}}, f_{d_{k+1}}). \tag{6.26}$$

(6.25) is the Bellman equation [18]. The optimal equations for the value function and the action value function are

$$V^*(e_{r_k}) = \min_{f_{d_k}} \left\{ r_{k+1} + \gamma V^*(e_{r_{k+1}}) \right\},$$

$$Q^*(e_{r_k}, f_{d_k}) = r_{k+1} + \gamma \min_{f_d} Q^*(e_{r_{k+1}}, f_d), \tag{6.27}$$

which satisfy (6.21) and (6.22).

The following inequality gives the condition if the control policy $h(e_{r_k})$ is applied

$$Q_{k+1}^h(e_{r_k}, f_{d_k}) < Q_k^h(e_{r_k}, f_{d_k}). \tag{6.28}$$

(6.28) means that if the action value function in the instance $k + 1$ is smaller than that in the instance k, then the new policy h is better than the previous one.

We have to use all the possible state-action pairs to obtain the optimal control policy. It is called greedy policy [15]. Greedy policies are impossible when the search space is huge.

We use the following recursive process to reduce the search space: We start from the policy h_1 and calculate its action value Q^{h_1}; then we choose another control policy h_2 and its action value Q^{h_2}. If (6.28) is correct, we have a better policy. If (6.28) is not satisfied, we continue with the recursive process until the optimal policy is found.

The Monte Carlo method is a simple estimation process, which calculates the average value function as

$$Q_{k+1}^h(e_{r_k}, f_{d_k}) = \left(1 - \frac{1}{k}\right) Q_k^h(e_{r_k}, f_{d_k}) + \frac{1}{k} R_k, \tag{6.29}$$

where $1/k$ is the learning rate, R_k is the reward after the current episode[1] has finished. For a non-stationary problem, (6.29) is rewritten as

$$Q_{k+1}^h(e_{r_k}, f_{d_k}) = \left(1 - \alpha_k\right) Q_k^h(e_{r_k}, f_{d_k}) + \alpha_k R_k, \tag{6.30}$$

where $\alpha_k \in (0, 1]$ is the learning rate.

The above Monte Carlo approximation is simple; however, its accuracy is poor. We can use the temporal difference (TD) learning methods, such as Q-learning and Sarsa, to get the optimal force $f_{d_k}^*$.

The basic idea is to use a real reward for the estimated value function Q in (6.30). Q-learning [18] is an off-policy algorithm that calculates the optimal Q-function in an iterative algorithm. The update rule is

$$Q_k^h(e_{r_k}, f_{d_k}) = Q_k^h(e_{r_k}, f_{d_k}) + \alpha_k \delta_k, \tag{6.31}$$

where the TD-error δ_k is

$$\delta_k = r_{k+1} + \gamma \min_{f_d} Q_k^h(e_{r_{k+1}}, f_d) - Q_k^h(e_{r_k}, f_{d_k}). \tag{6.32}$$

To obtain the optimal Q-function, large exploration of the state-action pair (e_{r_k}, f_{d_k}) is required. Algorithm 6.1 gives the Q-learning algorithm for the optimal desired force.

Sarsa (State-action-reward-state-action) is an on-policy algorithm of TD-learning. The major differences between Sarsa and Q-learning are that for Sarsa, the minimum Q of the next state is not needed for the Q-values update, and a new action is selected using the same policy. The Sarsa update rule is the same as (6.31), but the TD-error δ_k is

$$\delta_k = r_{k+1} + \gamma Q_k^h(e_{r_{k+1}}, f_{d_{k+1}}) - Q_k^h(e_{r_k}, f_{d_k}). \tag{6.33}$$

Algorithm 6.2 shows the Sarsa learning algorithm.

---

1 An episode is a sequence of interactions between initial and terminal states or a certain number of steps.

Algorithm 6.1 Q learning for the impedance model

1: **for** every episode **do**
2:     Initialize e_{r_0} and Q_0
3:     **repeat**
4:       **for** every time step $k = 0, 1, \ldots$ **do**
5:         Take action f_{d_k}
6:         Apply f_{d_k}, measure next state $e_{r_{k+1}}$ and reward r_{k+1}
7:         Update Q_k^h with (6.31) and (6.32)
8:         $e_{r_k} \leftarrow e_{r_{k+1}}$
9:       **end for**
10:    **until** terminal state
11: **end for**

Algorithm 6.2 Sarsa algorithm for the impedance model

1: **for** every episode **do**
2:     Initialize e_{r_0} and Q_0
3:     **repeat**
4:       **for** every time step $k = 0, 1, \ldots$ **do**
5:         Take action f_{d_k}
6:         Apply f_{d_k}, measure next state $e_{r_{k+1}}$ and reward r_{k+1}
7:         Update Q_k^h with (6.31) and (6.33)
8:         $e_{r_k} \leftarrow e_{r_{k+1}}$
9:         $f_{d_k} \leftarrow f_{d_{k+1}}$
10:      **end for**
11:    **until** terminal state
12: **end for**

To accelerate learning convergence, Q-learning and Sarsa algorithms are modified using eligibility traces [18–20], which offer a better way to assign credits for visited states. The eligibility trace for the pair (e_r, f_d) at time step k is

$$e_k(e_r, f_d) = \begin{cases} 1, & e_r = e_{r_k}, f_d = f_{d_k} \\ \lambda \gamma e_{k-1}(e_r, f_d), & \text{otherwise.} \end{cases} \tag{6.34}$$

It decays with time by the factor $\lambda \in [0, 1)$. So recently visited states have more eligibility for receiving credit.

The $Q(\lambda)$ and Sarsa(λ) algorithms have the same structure as in (6.31), where the optimal Q value function is calculated by

$$Q_k^h(e_{r_k}, f_{d_k}) = Q_k^h(e_{r_k}, f_{d_k}) + \alpha_k \delta_k e_k(e_r, f_d). \tag{6.35}$$

The Sarsa algorithm takes into account the control policy and incorporates with its action values. The Q-learning simply assumes that an optimal policy is followed.

In the Sarsa algorithm, the policy for the value function V and the policy for the action value function Q are the same. In Q-learning, the policies for V and Q are different. The minimum of $Q^h(e_{r_{k+1}}, f_{d_k})$ is applied to update $Q^h(e_{r_k}, f_{d_k})$. This means the minimum action does not have relation with $f_{d_{k+1}}$. To obtain the $f_{d_k}^*$, the Sarsa algorithm is better because it uses the improved policies. The following theorems give the convergence of our complete approach.

Theorem 6.1 *The reinforcement learning (6.31) for the desired force in an unknown environment converges to the optimal value f_d^* if*

1) *The state and action spaces are finite.*
2) *$\sum_k \alpha_k = \infty$ and $\sum_k \alpha_k^2 < \infty$ uniformly over the state-action pair w.p. 1.*
3) *The variance is finite.*

The proof can be found in [21]. Since $0 < \alpha_k \leq 1$, the second condition requires that all state-action pairs be visited infinitely often. Under the condition of the convergence of the reinforcement learning, the following theorem gives the stability of the closed-loop system.

Theorem 6.2 *If the robot (A.17) is controlled by the PID control (6.5) and the admittance controller (6.9), then the closed-loop system is semi-global asymptotically stable if the PID control gains satisfy*

$$\lambda_{\min}(K_p) \geq \frac{3}{2}k_g,$$

$$\lambda_{\max}(K_i) \leq \phi \frac{\lambda_{\min}(K_p)}{\beta},$$

$$\lambda_{\min}(K_v) \geq \phi + \beta, \tag{6.36}$$

where $\phi = \frac{\lambda_{\min}(M(q))\lambda_{\min}(K_p)}{3}$, and the contact force f_e satisfies

$$B_a, K_a > 0,$$

$$x_e \leq x_r \leq x_d,$$

$$\|e_f\| \leq C\|x_e - x_d\|, \tag{6.37}$$

for some $C > 0$.

Proof: The proof of the PID controller conditions are given in [6, 22]. We only give the proof of the admittance controller. The first condition of (6.37) is satisfied due the admittance controller design. The second condition of (6.37) considers the environment dynamics (6.1) and the desired force (6.9) using the PID controller, i.e., $x = x_r$:

$$f_e = -C_e \dot{x}_r + K_e(x_e - x_r), \tag{6.38}$$

$$f_d = L(x_r - x_d). \tag{6.39}$$

The force error between (6.38) and (6.39) is

$$
\begin{aligned}
e_f &= f_e - f_d \\
&= -C_e \dot{x}_r + K_e x_e - (K_e + L)x_r + L x_d.
\end{aligned} \tag{6.40}
$$

Substituting (6.40) into (6.9) yields

$$x_r(s) = \left(sI + \Lambda^{-1}\Theta\right)^{-1} \left[x_r(0) + \Lambda^{-1}K_e x_e(s) + \Lambda^{-1}(K_a + L)x_d(s)\right],$$

where $\Lambda = B_a + C_e$, and $\Theta = K_e + K_a + L$. The solution is

$$
\begin{aligned}
x_r(t) &= \exp^{-\Psi t} x_r(0) + \Psi^{-1}\left[\left(I - \exp^{-\Psi t}\right)\Lambda^{-1}K_e x_e \right. \\
&\quad \left. + \left(I - \exp^{-\Psi t}\right)\Lambda^{-1}(K_a + L)x_d\right],
\end{aligned} \tag{6.41}
$$

where $\Psi = \Lambda^{-1}\Theta$. When $t \to 0$, we have

$$x_r(0) = x_r(0). \tag{6.42}$$

When $t \to \infty$, we have

$$x_r(\infty) = \Psi^{-1}\left[\Lambda^{-1}K_e x_e + \Lambda^{-1}(K_a + L)x_d\right]. \tag{6.43}$$

Since the environment dynamics and the impedance model are assumed decoupled, the position reference (6.43) can be simplified as

$$x_r(\infty) = \Theta^{-1}\left[K_e x_e + (K_a + L)x_d\right]. \tag{6.44}$$

Note that the damping of the admittance model does not appear in (6.44); however, it changes the response of x_r. When $K_e \gg K_a + L, x_r(\infty) \approx x_e$. This means the environment is stiffer than the robot end effector. When $K_a + L \gg K_e, x_r(\infty) \approx x_d$, which means the stiffness of the end effector is bigger than the environment's stiffness. So the second condition of (6.37) is correct. To prove the third condition of (6.37), let us suppose the following cases:

1) When $K_e \gg K_a + L, x_r(\infty) \approx x_e$, the force error is

$$e_f = -L(x_e - x_d) \le \lambda_{\max}(L)\|x_e - x_d\|.$$

2) When $K_e \ll K_a + L, x_r(\infty) \approx x_d$,

$$e_f = K_e(x_e - x_d) \le \lambda_{\max}(K_e)\|x_e - x_d\|.$$

3) When $K_e = K_a + L$,

$$e_f = \frac{1}{2}(K_e - L)(x_e - x_d) \le \frac{1}{2}\lambda_{\max}(K_e - L)\|x_e - x_d\|.$$

4) When $K_e = K_a = L$,

$$e_f = \frac{1}{3}K_e(x_e - x_d) \le \frac{1}{3}\lambda_{\max}(K_e)\|x_e - x_d\|.$$

The above cases satisfy the third condition of (6.37), which completes the proof. ∎

The force error is zero when the desired position x_d is equal to the position of the environment, x_e. The admittance parameters must be selected according to the environment. If the environment is rigid, it is preferable to choose the admittance parameters small enough so that x_r is far from x_d. If the environment is not so rigid, large values of the admittance parameters can lead to a reference position x_r near x_d.

6.4 Simulations and Experiments

Example 6.1 *Simulations*

First we use simulations to evaluate the reinforcement learning controllers performance. We propose the following parameters for the dynamics of the environment (6.2): $A_e = -5$, $B_e = -0.005$. The desired Cartesian position is $x_d = [0.0495\ 0\ 0.1446]^T$. Since the robot has contact with the environment, the desired position in the X-axis is greater than the environment position x_e. The weights of the reward (or utility function) are $S = 1$ and $R = 0.1$. The solution of (6.11) with (6.12) is $L = 960$; then the desired force is

$$f_{d_k}^* = L(X_k - X_{d_k}) = 960(X_k - X_{d_k}),\tag{6.45}$$

where X_k is the position of the robot in the X-axis, and X_{d_k} is the desired position in the X-axis. This real solution is used to compare it with the solutions obtained using the Q-learning and Sarsa algorithms.

In order to select the parameters of the reinforcement learning, we use a random search method. The prior knowledge is applied to give the parameter range. It is similar as the ε-greedy exploration strategy [10]. The learning parameters are given in Table 6.1.

In each condition, there are 1000 episodes, and each episode has 1000 training steps. We compare the four reinforcement learning methods given in this chapter: Q-learning (algorithm 6.1), Sarsa learning (algorithm 6.2), $Q(\lambda)$ as defined in (6.35), and Sarsa(λ) as defined in (6.35).

Figure 6.3 gives the reinforcement learning results. We can see that all learning methods converge to the optimal solution of LQR. By the admittance control method, the contact force almost converges to the solution of LQR.

Table 6.1 Learning parameters

Parameter	Description	Value
α	Learning rate	0.9
γ	Discount factor	0.9
λ	Trace decay factor	0.65
ε	ε-greedy probability	0.1
ε-decay	ε-greedy decay factor	0.99

Finally, we use the PID controller to track the virtual reference x_r, which is obtained from the desired force. When there is no contact force, the position x follows x_d; on the other hand, when the robot touches the environment, the desired position x_d is modified by the admittance control and the robot position follows the new reference x_r. We can see that $Q(\lambda)$ converges to the solution in fewer episodes than the Sarsa algorithm.

If the environment parameters are changed to $A_e = -4$ and $B_e = -0.002$ and also the desired position is changed to $x_d = [0.0606, 0, 0.1301]^T$, the reinforcement learning methods can learn the environment automatically. The learning curves are shown in Figure 6.4. In this new environment, the desired force is

$$f_{d_k}^* = L(X_k - X_{d_k}) = 1875(X_k - X_{d_k}). \tag{6.46}$$

RL has less overshoots because the initial Cartesian position is near the desired position. Another advantage of reinforcement learning is its smooth learning process. After long-term learning, the desired force converges very fast.

Example 6.2 *Experiments*

To evaluate the reinforcement learning methods proposed in this chapter, we use a 2-DOF pan and tilt robot, shown in Figure 6.5, and a 6-DOF force/torque (F/T) sensor, which is mounted on the end effector. The environment is a steel box with unknown stiffness and damping. The real-time control environment is Simulink and MATLAB 2012. The initial conditions of the joint angles are $q_1(0) = 0$ and $q_2(0) = \frac{\pi}{2} - 0.02$, which avoid the Jacobian singularity. These joint angles correspond to the Cartesian position $x(0) = [0.014\ 0\ 0.1651]^T$.

In the experiment, the environment is located on the X-axis at the position $X_e = 0.045$ m. To calculate the the optimal solution, we need the parameters of the environment, i.e., the damping C_e and the stiffness K_e. We use the recursive least squares (RLS) method (2.30) to estimate them. The force can be expressed as

$$f_e = \varphi^T \theta,$$

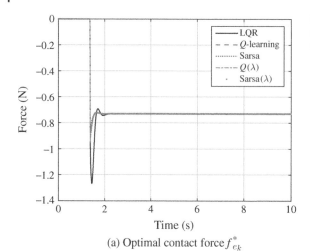

Figure 6.3 Position/force control with $A_e = -5$, $B_e = -0.005$

(a) Optimal contact force $f_{e_k}^*$

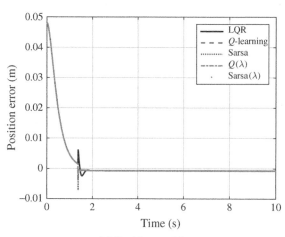

(b) Position tracking error.

where $\varphi = [-\dot{x}, -x]^\mathsf{T}$ is the regression vector, and $\theta = [C_e, K_e]^\mathsf{T}$ is the parameter vector. The parameter estimation is shown in Figure 6.6.

The parameters converge to the following values: $\hat{K}_e \approx 2721.4$ N/m and $\hat{C}_e \approx 504.0818$ Ns/m. We rewrite them in the form of (6.2); then $A_e = -4.3987$ and $B_e = -0.002$. If we choose $S = 1$ and $R = 0.1$, the solution of the DARE (6.11) with the above environment parameters is

$$f_{d_k} = L(X_k - X_{d_k}) = 2606.7(X_k - X_{d_k}). \tag{6.47}$$

In the experiment, we use the same learning parameters as in Table 6.1. Similar as the random search method for Table 6.1, the proposed gains of the PID

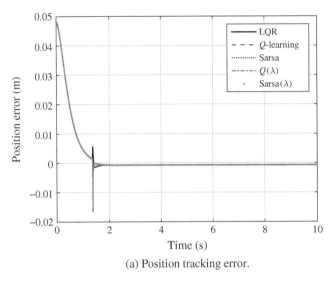

(a) Position tracking error.

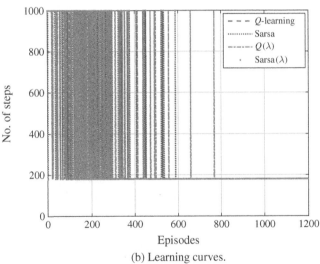

(b) Learning curves.

Figure 6.4 Position/force control with $A_e = -4$ and $B_e = -0.002$

and admittance controllers are given in Table 6.2. Due to the high friction of the robot, we do not need the derivative gains. We apply 30 real-time experiments, each one lasting 10 seconds. The experiment results are given in Figure 6.7. The contact force obtained from the reinforcement learning converges to the LQR solution.

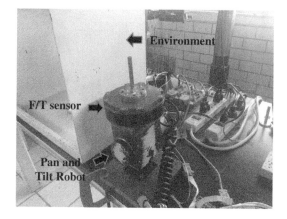

Figure 6.5 Experimental setup

Table 6.2 Controllers gains

Gain	Description	Value
K_p	Proportional Gain	$50I_{3\times3}$
K_d	Derivative Gain	$0I_{3\times3}$
K_i	Integral Gain	$10I_{3\times3}$
B_a	Damping	500
K_a	Stiffness	2500

Discussion

From the above results, both LQR and RL converge to the optimum force $f^*_{d_k}$. For LQR, the knowledge of the environment parameters (6.2) is mandatory to obtain the DARE solution (6.11), and the parameters can be solved by any identification algorithm, such as the RLS algorithm. These methods need three steps: (1) the identification of the environment, (2) the solution of the DARE using the environmental estimates, (3) the completion of the experiment for the position/force controller.

The RL algorithms do not require knowledge of the environment dynamics. The success of the algorithms lies in the exploration of all the state-action pairs. The learning and control procedure is a one-step experiment, and the weakness of these algorithms is the learning time and the selection of the learning parameters to assure the convergence of the solution.

Another advantage of using RL is that we can avoid large errors in transient time. The LQR does not need the contact force between the robot and the environment. If the initial conditions of the robot are close to the desired position, the initial

Figure 6.6 Environment estimation

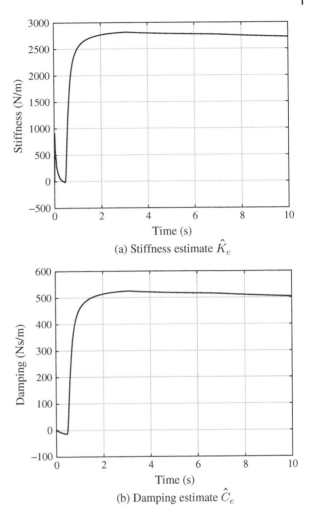

(a) Stiffness estimate $\hat{K}_e$

(b) Damping estimate $\hat{C}_e$

position error will be small and the desired force in the transient time will also be small. The RL solution in the transient time is close to zero due to two factors: the noise of the force sensor and the absence of contact force. The RL algorithm avoids abruptly changing the desired force from zero to the desired force.

The hyper-parameters are important for the algorithm convergence because we need to explore new policies and exploit our current knowledge. If the discount factor γ is too small, the RL method will converge fast by exploiting its current knowledge but will not converge to the optimal solution. If the discount factor is near 1, the RL method will converge slowly to the optimal solution by exploring the

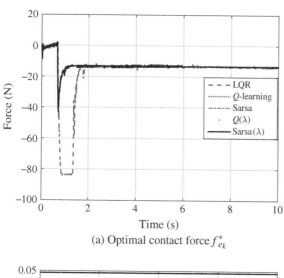

Figure 6.7 Experiment results

(a) Optimal contact force $f_{e_k}^*$

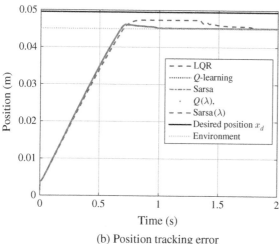

(b) Position tracking error

whole space. The other hyper-parameters are chosen according to the experiment performance.

The main disadvantage of using RL methods is the greedy actions, since one needs to perform an optimization procedure in each state to find the action that leads to an optimal value. This can be computationally intensive, especially if the action space is continuous. Therefore, the RL methods generally discretize the action space, and the optimization over the action space becomes a matter of enumeration.

6.5 Conclusions

In this chapter we address a new robot-interaction control scheme with RL for the position/force control. Unlike the previous robot-interaction control schemes, this controller uses both impedance and admittance control law to achieve force and position tracking. The environment parameters are unknown. We use reinforcement learning to learn the optimal desired force, which is equivalent to the optimal impedance model, and then an admittance control law guarantees both force and position tracking using an inner position control law. Simulations and experiments are carried out to verify the approach with satisfactory results using a 2-DOF pan and tilt robot and an F/T sensor.

References

1 S. Chiaverini, B. Siciliano, and L. Villani, "A Survey of Robot Interaction Control Schemes with Experimental Comparison," *IEEE/ASME Transactions on Mechatronics*, vol. 4, pp. 273–285, September 1999.

2 M. Tufail and C. de Silva, "Impedance Control Schemes for Bilateral Teleoperation," *International Conference on Computer Science and Education*, pp. 44–49, August 2014. Vancouver, Canada.

3 N. Hogan, "Impedance Control: An Approach to Manipulation," *Journal of Dynamic Systems, Measurement, and Control*, vol. 107, pp. 1–24, March 1985. Transactions of the ASME.

4 G. Ferreti, G. Magnani, and P. Rocco, "Impedance Control for Elastic Joints Industrial Manipulators," *IEEE Transactions on Robotics and Automation*, vol. 20, pp. 488–498, June 2004.

5 S. Singh and D. Popa, "An Analysis of Some Fundamental Problems in Adaptive Control of Force and Impedance Behavior: Theory and Experiments," *IEEE Transactions on Robotics and Automation*, vol. 11, pp. 912–921, December 1995.

6 W. Yu, J. Rosen, and X. Li, "PID Admittance Control for an Upper Limb Exoskeleton," *American Control Conference*, pp. 1124–1129, June 29–July 01 2011. O'Farrel Street, San Francisco, California.

7 R. Bonitz and T. Hsia, "Internal Force-Based Impedance Control for Cooperating Manipulators," *IEEE Transactions on Robotics and Automation*, vol. 12, pp. 78–89, February 1996.

8 R. Johansson and M. Spong, "Quadratic optimization of impedance control," *Proc. IEEE International Conference in Robot Autom*, no. 1, pp. 616–621, 1994.

9 M. Matinfar and K. Hashtrudi-Zaad, "Optimization-based robot compliance control: geometric and linear quadratic approaches," *Int. J. Robot. Res.*, vol. 24, no. 8, pp. 645–656, 2005.

10 C. Wang, Y. Li, S. Sam Ge, and T. Heng Lee, "Optimal critic learning for robot control in time-varying environments," *IEEE Transactions on Neural Networks and Learning Systems*, vol. 26, no. 10, 2015.

11 J. Wen and S. Murphy, "Stability analysis of position and force control for robot arms," *IEEE Transactions on Automatic Control*, vol. 36, no. 3, pp. 365–371, 1991.

12 L. Wilfinger, J. Wen, and S. Murphy, "Integral force control with robustness enhancement," *IEEE Contr. Syst. Mag*, vol. 14, pp. 31–40, 1994.

13 A. Perrusquía, W. Yu, and A. Soria, "Position/force control of robots manipulators using reinforcement learning," *Industrial Robot: the international journal of robotics research and application*, vol. 46, no. 2, pp. 267–280, 2019.

14 A. Khan, D. Yun, M. Ali, J. Han, K. Shin, and C. Han, "Adaptive Impedance Control for Upper Limb Assist Exoskeleton," *IEEE International Conference on Robotics and Automation*, pp. 4359–4366, May 26–30 2015. Seattle, Washington.

15 F. Lewis, D. Vrable, and K. Vamvoudakis, "Reinforcement learning and feedback control using natural decision methods to desgin optimal adaptive controllers," *IEEE Control Systems Magazine*, 2012.

16 M. Dimeas and N. Aspragathos, "Online stability in human-robot cooperations with admittance control," *IEEE Transactions on Haptics*, vol. 9, no. 2, pp. 267–278, 2016.

17 A. Perrusquía and W. Yu, "Task space human-robot interaction using angular velocity jacobian," *2019 International Symposium in Medical Robotics (ISMR), Atlanta, Georgia*, pp. 1–7, 2019.

18 R. Sutton and B. A, *Reinforcement Learning: An Introduction*. Cambridge, MA: MIT Press, 1998.

19 I. Grondman, M. Vaandrager, L. Buşoniu, R. Babûska, and E. Schuitema, "Efficient model learning methods for actor-critic control," *IEEE Transactions on Systems, man, and cybernetics. Part B: Cybernetics*, vol. 42, no. 3, 2012a.

20 I. Grondman, L. Buşoniu, and R. Babûska, "Model learning actor-critic algorithms: performance evaluation in a motion control task," *51st IEEE Conference on Decision and Control (CDC)*, pp. 5272–5277, 2012b.

21 T. Jaakola, J. M., and S. Singh, "On the convergence of stochastic iterative dyanamic programming algorithms," *Neural Computation*, vol. 6, no. 6, pp. 1185–1201, 1994.

22 W. Yu, R. Carmona Rodriguez, and X. Li, "Neural PID Admittance Control of a Robot," *American Control Conference*, pp. 4963–4968, June 17–19 2013. Washington, DC.

7

Continuous-Time Reinforcement Learning for Force Control

7.1 Introduction

Q-value function approximation is a big topic in reinforcement learning (RL). When the learning problem has very large state and action space, it becomes necessary to use function approximations methods to represent the value functions [1]. Some of the most popular basis functions are the Fourier function, Gaussian kernels, and local linear regression [2, 3].

Gaussian kernels are non-parametric approximators that have certain smoothness properties and are useful for approximating continuous functions. These functions are localized functions. It is well known that parametric approximators are better than non-parametric ones [2], because of convergence analysis and implementation. The parametric form of the Gaussian kernels are the radial basis functions (RBFs), where one input state is fixed as the center of the RBF. It is a tedious procedure when the number of RBFs increases for large input data [3].

A simple way to organize the input data into groupings is the clustering method, which is an unsupervised learning method [4]. There exist several references that use clustering to design efficient feature spaces in order to estimate optimal policies [5–9]. These online clustering provides adaptive clusters while they are learning. They require enough exploration of the input space to obtain good clusters, and the obtained clusters cover only the visited input state. Off-line clustering avoids this problem [5, 8].

To apply RL in continuous time, the input space has to be divided into many cells. The computation becomes complex, and the accuracy is decreased. Some approximations [10] are used to reduce the computation space such as fuzzy representations [11], the backward Euler approximation [12], Taylor series [13–15],

Human-Robot Interaction Control Using Reinforcement Learning, First Edition. Wen Yu and Adolfo Perrusquía.
© 2022 The Institute of Electrical and Electronics Engineers, Inc. Published 2022 by John Wiley & Sons, Inc.

and parametric and non-parametric approximators [2, 3], such as Gaussian kernels. The above methods have good results; however, they require an exhaustive exploration phase to find the near-optimal solution.

In this chapter, we discuss the large and continuous space problems in RL and apply them in the position/force control. A novel hybrid solution is given to take both advantages of discrete-time and continuous-time RLs.

7.2 K-means Clustering for Reinforcement Learning

Since we are working with a large state-action space, an approximation is required. We normalize radial basis functions (NRBFs) as the approximator of the Q-value function. The parametrization is given by

$$\hat{Q}_k(e_{r_k}, f_{d_k}) = \sum_{i=1}^{n} \phi_i(e_{r_k}, f_{d_k})\theta_i = \Phi^{\mathsf{T}}(e_{r_k}, f_{d_k})\theta = F(\theta), \tag{7.1}$$

where $\Phi(e_{r_k}, f_{d_k})$ is the column vector of the normalized RBFs:

$$\phi_i(e_r, f_d) = \frac{\exp\left(-\frac{1}{2}\begin{bmatrix} e_r - c_i \\ f_d - c_i \end{bmatrix}^{\mathsf{T}} \beta_i^{-1} \begin{bmatrix} e_r - c_i \\ f_d - c_i \end{bmatrix}\right)}{\sum_{i=1}^{D} \exp\left(-\frac{1}{2}\begin{bmatrix} e_r - c_i \\ f_d - c_i \end{bmatrix}^{\mathsf{T}} \beta_i^{-1} \begin{bmatrix} e_r - c_i \\ f_d - c_i \end{bmatrix}\right)}, \tag{7.2}$$

were the vector $c_i = [c_{i,1}, \dots, c_{i,D}]^{\mathsf{T}} \in \mathbb{R}^D$ is the center of the ith RBF, the symmetric positive-definite matrix $\beta_i \in \mathbb{R}^{D \times D}$ is the width, and D is the number of RBFs.

The main problem of using NRBFs is the number of RBFs and the location of the centers. In [3], the centers of the RBFs are fixed and close to the equilibrium point. This method helps to stabilize the closed-loop system, such as pendubot, acrobat, and cart pole systems. The number of RBFs depends on the dimension of the input state. Few RBFs may cause a wrong estimation. Increasing the number of RBFs may obtain a good estimation, but it can present over-fitting problems.

One method to obtain the centers of the RBFs is to use K-means clustering to partition the input state into K clusters. The K-means clustering algorithm gives the partition of the set such that the squared error between the mean of the cluster and all points within the cluster are minimized. The squared error of the mean of the cluster c_j is defined as

$$\mathcal{J}(c_j) = \sum_{y_i \in c_j} \|y_i - \mu_j\|^2.$$

The main goal of K-means is to minimize the sum of the quadratic error over all the K clusters,

$$J(C) = \underset{C}{\operatorname{argmin}} \sum_{j=1}^{K} \sum_{y_i \in c_j} \|y_i - \mu_j\|^2. \tag{7.3}$$

The number of clusters is the number of RBFs of the approximator. Let $\mathcal{Y} = \{y_i\}, i = 1, \dots, n'$, be the set of n'-dimensional points to be clustered into a set of K clusters, n' is the number of points of the input, and $C = \{c_j, j = 1, \dots, K\}$. Each centroid of a cluster is a collection of feature values that define the resulting groups.

The K-means algorithm is given in Algorithm 7.1.

Algorithm 7.1 K-means Clustering

1: **Input:** K,
2: Initialize the cluster centroids $C = \{c_j, j = 1, \dots K\}$,
3: **repeat**
4:     Find the nearest centroid for each point, $I_j \leftarrow \operatorname{argmin}_{c_j \in C} |y - c_j|^2$,
5:     Update the centroids $c_j \leftarrow \frac{1}{|I_j|} \sum_{y_i \in I_j} y_i$
6: **until** the clusters memberships are stabilized

The algorithm gets K random centers of the approximators using NRBFs. The RL approximation algorithm is based in a gradient descent method that minimizes the squared error between the optimal value (target learning) and the current Q value:

$$\theta_{k+1} = \theta_k + \alpha_k \left[Q^* - \hat{Q}_k(e_{r_k}, f_{d_k}) \right] \Phi(e_{r_k}, f_{d_k}).$$

Substituting the parametrization (7.1) into above equation and modifying the target learning with its approximation, we obtain the Sarsa update rule as

$$\theta_{k+1} = \theta_k + \alpha_k \delta_k \Phi(e_{r_k}, f_{d_k}). \tag{7.4}$$

Here the TD-error for both Q-learning and Sarsa approximation is given as follows:

$$Q - learning : \delta_k = r_{k+1} + \gamma \underset{f_d}{\min} \left(\Phi^\top(e_{r_{k+1}}, f_d)\theta_k \right) - \Phi^\top(e_{r_k}, f_{d_k})\theta_k$$

$$Sarsa : \delta_k = r_{k+1} + \gamma \Phi^\top(e_{r_{k+1}}, f_{d_{k+1}})\theta_k - \Phi^\top(e_{r_k}, f_{d_k})\theta_k. \tag{7.5}$$

Both algorithms require enough exploration for reliable approximation. For a policy iteration, the optimal policy is given by

$$h^*(e_r) \in \underset{f_d}{\operatorname{arg\,min}} F(\theta^*) = \underset{u}{\operatorname{argmin}} \hat{Q}(e_r, f_d).$$

The convergence of the algorithm lies in the following contraction property:

$$\|(F^\dagger \circ \mathcal{H} \circ F)(\theta) - (F^\dagger \circ \mathcal{H} \circ F)(\theta')\| \leq \gamma' \|\theta - \theta'\|,$$

where $F^\dagger$ is the pseudo-inverse projection of F, $\mathcal{H}$ is the Bellman operator, and γ' is the contraction constant, which is equivalent to a discount factor. To prove the convergence, the following assumptions are needed.

A7.1 The learning rate for the parameter updating law (7.4) satisfies

$$\sum_k^\infty \alpha_k = \infty, \qquad \sum_k^\infty \alpha_k^2 < \infty.$$

A7.2 The basis function satisfies

$$\sum_{i=1}^K |\phi_i(x, u)| \leq 1.$$

A7.3 Let $\mathcal{H}$ be the mapping from the space of Q functions to itself,

$$\|\mathcal{H}(Q) - \mathcal{H}(Q')\| \leq \gamma \|Q - Q'\|,$$

then $\mathcal{H}$ is a contraction with respect to $\|\cdot\|$.

A7.4 There exists a mapping $\mathcal{H}' = F^\dagger \circ \mathcal{H} \circ F$ with contraction constant $\gamma' = \gamma$, such that

$$\theta = F^\dagger(\mathcal{H}'(F(\theta))),$$

where θ, θ' and Q, Q' satisfy

$$\|F(\theta) - F(\theta')\| \leq \|\theta - \theta'\|,$$
$$\|F^\dagger(Q) - F^\dagger(Q')\| \leq \|Q - Q'\|.$$

A7.5 The variance is bounded as follows

$$var[\eta_k(x)|F_k] \leq \zeta \left(1 + \|F(\theta^*) - Q^*\|^2\right),$$

where η_k is a noise term, F_k is a sequence of increasing σ-fields, and ζ is a positive constant.

Assumption A7.1 requires that all state-action pairs are visited infinitely often. A7.2-A7.4 are easy to prove by means of the contraction property, the boundedness of the basis functions, and considering that A7.5 is a consequence of assumptions A7.1-A7.4.

Theorem 7.1 *If assumptions A7.1-A7.5 hold and the mappings $\mathcal{H}$ and $\mathcal{H}'$ are a contraction with respect to the maximum norm, the update rule (7.4) converges to the unique fixed point θ^* with probability 1 and*

$$\|Q^* - F(\theta^*)\| \leq \frac{2}{1 - \gamma} e,$$

$$\|Q^* - \hat{Q}^{h^*}\| \le \frac{4\gamma}{(1-\gamma)^2}e,$$

where $e = \min_\theta \|Q^* - F(\theta)\|$.

Proof: Using Assumptions A7.3 and A7.4, we have

$$\|\theta - \mathcal{H}'(\theta)\| \le \|F^\dagger(F(\theta)) - F^\dagger(\mathcal{H}(F(\theta)))\|$$
$$\le \|F(\theta) - \mathcal{H}(F(\theta))\|$$
$$\le \|F(\theta) - Q^*\| + \|Q^* - \mathcal{H}(F(\theta))\|$$
$$\le e + \gamma e.$$

Using the above result on the optimal θ^*, we obtain

$$\|\theta^* - \theta\| \le \|\theta^* - \mathcal{H}'(\theta)\| + \|\mathcal{H}'(\theta) - \theta\|$$
$$\le \gamma\|\theta^* - \theta\| + (1+\gamma)e \le \frac{1+\gamma}{1-\gamma}e.$$

For the Q-value iteration

$$\|Q^* - F(\theta^*)\| \le \|Q^* - F(\theta)\| + \|F(\theta) - F(\theta^*)\|$$
$$\le e + \|\theta - \theta^*\|$$
$$\le e + \frac{1+\gamma}{1-\gamma}e \le \frac{2}{1-\gamma}e.$$

For the policy iteration algorithm

$$\|Q^* - \hat{Q}^{h^*}\| \le \|Q^* - \mathcal{H}(F(\theta^*))\| + \|\mathcal{H}(F(\theta^*)) - \hat{Q}^{h^*}\|$$
$$\le \gamma\|Q^* - F(\theta^*)\| + \gamma\|F(\theta^*) - \hat{Q}^{h^*}\|$$
$$\le 2\gamma\|Q^* - F(\theta^*)\| + \gamma\|Q^* - \hat{Q}^{h^*}\|$$
$$\le \frac{4\gamma}{(1-\gamma)^2}e.$$

Since we already proves that $\mathcal{H}'$ is a contraction, the update rule converges to the unique fixed point θ^* with probability one. ∎

The update rule (7.4) is an one-step backup, whereas the reward needs a series of steps. Eligibility traces [3, 16] offer a better method to assign the credit to the visited states in several steps ahead. The eligibility trace for the state e_r at time step k is

$$e_k(e_r) = \begin{cases} 1, & \text{if } e_r = e_{r_k} \\ \lambda\gamma e_{k-1}(e_r), & \text{otherwise.} \end{cases} \tag{7.6}$$

It decays with time by a factor $\lambda\gamma$, with $\lambda\gamma \in [0,1)$. The states are more eligible for receiving credit. The update rules (7.4) is modified as

$$\theta_{k+1} = \theta_k + \alpha\delta_k \sum_{e_v \in \mathcal{E}_v} \left.\frac{\partial \hat{Q}_k(e_{r_k}, f_{d_k})}{\partial \theta_k}\right|_{\substack{e_r = e_v \\ \theta = \theta_{k-1}}} e_k(e_v), \tag{7.7}$$

where $\mathcal{E}_v$ is the set of states visited during the current episode.

7.3 Position/Force Control Using Reinforcement Learning

In the continuous-time case, the robot-environment interaction model (6.1) is expressed as

$$\dot{x}(t) = Ax(t) + Bf_e(t), \; x(t) \geq x_e, \tag{7.8}$$

where $A = -C_e^{-1}K_e$, and $B = -C_e^{-1}$. The impedance model is

$$\dot{e}_r(t) = Ae_r(t) + Bf_d(t). \tag{7.9}$$

The PID control in task space is (6.5), and the admittance controller is (6.8).

Reinforcement Learning in Continuous Time

As in (6.9), we want to design an optimal control as

$$f_d^*(t) = Le_r(t), \tag{7.10}$$

which minimizes the following continuous-time discounted cost function

$$V(e_r(t)) = J(e_r(t), u) = \int_t^\infty \left(e_r^\mathsf{T}(\tau)Se_r(\tau) + f_d^\mathsf{T}(\tau)Rf_d(\tau)\right)e^{-\gamma(\tau-t)}d\tau. \tag{7.11}$$

Applying the time derivative of (7.11) and Leibniz rule,

$$\dot{V}(e_r(t)) = \int_t^\infty \frac{\partial}{\partial t}r(\tau)e^{-\gamma(\tau-t)}d\tau - e_r^\mathsf{T}(t)Se_r(t) - f_d^\mathsf{T}(t)Rf_d(t), \tag{7.12}$$

where $r(\tau) = e_r(\tau)^\mathsf{T}Se_r(\tau) + f_d^\mathsf{T}(\tau)Rf_d(\tau)$ is the reward function. The optimal value function is

$$V^*(e_r(t)) = \min_{f_d} \int_t^\infty \left(e_r^\mathsf{T}(\tau)Se_r(\tau) + f_d^\mathsf{T}(\tau)Rf_d(\tau)\right)e^{-\gamma(\tau-t)}d\tau.$$

Using Bellman's optimality principle, we have the following Hamilton-Jacobi-Bellman (HJB) equation

$$\min_{f_d}\left\{\dot{V}^*(e_r(t)) + r(t) - \int_t^\infty \frac{\partial}{\partial t}r(\tau)e^{-\gamma(\tau-t)}d\tau\right\} = 0.$$

If the optimal desired force exists, the optimal value function is quadratic in the error $e_r(t)$:

$$V^*(e_r(t)) = e_r^\top(t) P e_r(t), \qquad (7.13)$$

where P satisfies the following differential Riccati equation:

$$e_r^\top(t)\left(A^\top P + PA + S - \gamma P\right) e_r(t) + 2e_r^\top(t) PB f_d(t) + f_d^\top(t) R f_d(t) = 0. \qquad (7.14)$$

By differentiating the Bellman equation (7.13), the optimal desired force is

$$f_d^*(t) = \left(-R^{-1} B^\top P\right) e_r(t) = L e_r(t). \qquad (7.15)$$

In order to obtain the existence condition of (7.14), we rewrite (7.15) as the optimal state-action form

$$Q^*(e_r(t), f_d(t)) = V^*(e_r(t)) = \begin{bmatrix} e_r(t) \\ f_d(t) \end{bmatrix}^\top H \begin{bmatrix} e_r(t) \\ f_d(t) \end{bmatrix}, \qquad (7.16)$$

where $Q\left[e_r(t), f_d(t)\right]$ is the value function

$$H = \{H_{ij}\} = \begin{bmatrix} \left(A - \frac{\gamma}{2}I\right)^\top P + P\left(A - \frac{\gamma}{2}I\right) + S & PB \\ B^\top P & R \end{bmatrix}.$$

Because $\partial Q^*/\partial f_d = 0$, the optimal control (7.15) is

$$f_d^*(t) = -H_{22}^{-1} H_{21} e_r(t).$$

There exists an unique positive solution P for (7.14) if the pair (A, B) is stable and the pair $(S^{1/2}, A)$ is observable. The above solution is the off-line method. In order to apply the online method, the Hewer algorithm or Lyapunov recursions algorithm can be used [17].

An equivalent form of the Bellman equation that does not require knowledge of the environment dynamics can be written as the integral reinforcement learning (IRL) as [14]

$$V(e_r(t)) = \int_t^{t+T} r(\tau) e^{-\gamma(\tau-t)} d\tau + e^{-\gamma T} V(e_r(t+T)), \qquad (7.17)$$

for any time $t \geq 0$ and $T > 0$.

For any fixed policy $f_d(t) = h(e_r(t))$, the following holds

$$V\left[e_r(t+T)\right] = Q\left(e_r(t+T), f_d(t+T)\right). \qquad (7.18)$$

Substituting (7.18) into (7.17), the Bellman equation of the Q-function is

$$Q\left[e_r(t), f_d(t)\right] = \int_t^{t+T} r(\tau) e^{-\gamma(\tau-t)} d\tau + e^{-\gamma T} Q\left(e_r(t+T), f_d(t+T)\right). \qquad (7.19)$$

The continuous-time reinforcement learning (7.19) needs the analytic solution of the integrator for all time. We use a discretization approximation method. The

Q-function update law (7.19) is computed by minimization of the following temporal difference error $\xi(t)$ as

$$\xi(t) = Q(e_r(t), f_d(t)) - \int_t^\infty r(\tau) e^{-\gamma(\tau-t)} d\tau.$$

The minimization of $\xi(t)$ is given by its time derivative as

$$\delta(t) = \dot{\xi}(t) = r(t) + \dot{Q}\left(e_r(t), f_d(t)\right) - \gamma Q\left(e_r(t), f_d(t)\right). \tag{7.20}$$

The expression (7.20) is the continuous-time counterpart of the discrete-time TD error (6.32). We use the backward Euler approximation [12] to approximate $\dot{Q}(\cdot)$ as

$$\dot{Q}(e_r(t), f_d(t)) = \frac{Q\left[e_r(t), f_d(t)\right] - Q\left[e_r(t-T), f_d(t-t)\right]}{T}.$$

So (7.20) becomes

$$\delta(t) = r(t) + \frac{1}{T}(1 - \gamma T)\left[Q(e_r(t), f_d(t)) - Q(e_r(t-T), f_d(t-T))\right]. \tag{7.21}$$

(7.21) can be written as

$$\delta(t) = r(t+T) + \frac{1}{T}(1 - \gamma T)\left[Q(e_r(t+T), f_d(t+T)) - Q(e_r(t), f_d(t))\right]. \tag{7.22}$$

The TD error $\delta(t)$ serves as the direction of the gradient. We still need to calculate the Q-function. In continuous state-action space, an approximation is required, such as the normalized radial basis functions (NRBFs) for the Q-value function as in (7.1).

Now we use the gradient descent method to update the parameters θ in the action-value function estimate $\hat{Q}(e_r, f_d; \theta)$. The objective is to minimize the squared error in (7.22),

$$E(t) = \frac{1}{2}\delta^2(t).$$

The gradient of the squared error with respect to the parameter vector θ is

$$\frac{\partial E(t)}{\partial \theta(t)} = \frac{\delta(t)}{T}\left[(1 - \gamma T)\frac{\partial \hat{Q}(e_r(t+T), f_d(t+T); \theta(t))}{\partial \theta(t)} - \frac{\partial \hat{Q}(e_r(t), f_d(t); \theta(t))}{\partial \theta(t)}\right]. \tag{7.23}$$

To simplify (7.23), we use a TD(0) algorithm, which uses only the previous estimate of the gradient:

$$\dot{\theta}(t) = -\alpha(t)\frac{\partial E(t)}{\partial \theta(t)} = \alpha(t)\delta(t)\frac{1}{T}\frac{\partial \hat{Q}(e_r(t), f_d(t); \theta(t))}{\partial \theta(t)}, \tag{7.24}$$

where $\alpha(t)$ is the learning rate.

The following theorem gives the convergence of the continuous-time reinforcement learning for the optimal desired force

$$\hat{f}_d(t) = \arg\min_{f_d} \hat{Q}(e_r, f_d; \theta) \approx \arg\min_{f_d} \Phi^T(e_r, f_d)\theta, \tag{7.25}$$

where θ is updated by (7.24), $\Phi^T(e_r, f_d)$ is obtained from RBFNN and the K-means algorithm, and $\delta(t)$ is calculated by (7.22).

Theorem 7.2   *If there exists the equilibrium point θ^* such that*

$$\min_{f_d} \hat{Q}(e_r, f_d; \theta) = \hat{Q}^*(e_r, f_d; \theta^*) = \Phi^T(e_r, f_d)\theta^*,$$

then (1) this equilibrium point is globally asymptotically stable, and (2) the learning rate for the parameter updating law (7.24) satisfies

$$\sum_t^\infty \alpha(t) = \infty, \quad \alpha(t) \in (0, 1], \quad \sum_t^\infty \alpha^2(t) < \infty, \tag{7.26}$$

The estimated Q-function $\hat{Q}(e_r, f_d; \theta)$ will converge to a near optimal value function $\hat{Q}^*(e_r, f_d; \theta^*)$. If (1) the equilibrium point is globally asymptotic stable such that

$$\hat{f}_d(t) = \arg\min_{f_d} \hat{Q}(e_r, f_d; \theta) \approx \arg\min_{f_d} \Phi^T(e_r, f_d)\theta,$$

and (2) the learning rate for the parameter updating law (7.24) satisfies

$$\sum_t^\infty \alpha(t) = \infty, \quad \alpha(t) \in (0, 1], \quad \sum_t^\infty \alpha^2(t) < \infty,$$

Proof: Consider the TD-error (7.22). The parameter vector θ of the continuous time reinforcement learning (7.24) is updated as

$$\dot{\theta} = \left(r(t + T) + \frac{1}{T}(1 - \gamma T)\Phi^T(t + T)\theta(t) - \frac{1}{T}\Phi^T(t)\theta(t) \right) \Phi(t)$$

$$= f(\theta(t)), \tag{7.27}$$

where $\Phi(t + T) = \Phi\left(e_r(t + T), f_d(t + T)\right)$ and $\Phi(t) = \Phi\left(e_r(t), f_d(t)\right)$. The equilibrium point θ^* of (7.27) should satisfy the following relation

$$\theta^* = F^\dagger \circ \mathcal{H} \circ F(\theta) = F^\dagger \circ \mathcal{H} \circ \hat{Q}\left(e_r, f_d; \theta\right), \tag{7.28}$$

where $\mathcal{H}$ is the Bellman operator, and $F^\dagger$ is the projection operator, which is defined as the pseudoinverse operation as

$$\left[F^\dagger \hat{Q}\right](e_r, f_d; \theta) = \left[\Phi(t)\Phi^T(t)\right]^{-1} \Phi(t)\hat{Q}(e_r, f_d; \theta).$$

The Bellman operator $\mathcal{H}$ is obtained from (7.19):

$$\left[\mathcal{H}Q\right](e_r(t), f_d(t)) = r(t + T) + \frac{1}{T}(1 - \gamma T)Q(e_r(t + T), f_d(t + T)). \tag{7.29}$$

To guarantee that the operator $\mathcal{H}$ is a contraction in θ, it is required that

$$\sum_{i=1}^{K} |\phi_i(e_r, f_d)| \leq 1. \tag{7.30}$$

Condition (7.30) is evident, because the basis functions are Gaussian functions. $f(\theta)$ in (7.27) can be written as

$$f(\theta(t)) = f_1(\theta(t)) + f_2(\theta(t)),$$

where $f_1 = \left(r(t+T) + \frac{1}{T}(1 - \gamma T)\Phi^{\mathsf{T}}(t+T)\theta \right) \Phi(t)$, and $f_2 = -\frac{1}{T}\Phi(t)\Phi^{\mathsf{T}}(t)\theta$. By the contraction property, for any θ_1 and θ_2, the following holds:

$$\|f_1(\theta_1) - f_1(\theta_2)\|_\infty \leq \frac{1 - \gamma T}{T} \|\theta_1 - \theta_2\|_\infty,$$

$$\|f_2(\theta_1) - f_2(\theta_2)\|_\infty \leq \frac{1}{T} \|\theta_1 - \theta_2\|_\infty.$$

Furthermore, the equilibrium point θ^* of (7.27) satisfies $f(\theta^*) = 0$. For any p-norm, the following holds:

$$\frac{d}{dt}\|\theta - \theta^*\|_p = \|\theta - \theta^*\|_p^{1-p} \sum_{i=1}^{K} (\theta(i) - \theta^*(i))^{p-1}$$

$$\cdot \left\{ [f_1(\theta)_i - f_1(\theta^*)_i] - [f_2(\theta)_i - f_2(\theta^*)_i] \right\}.$$

Using Hölder's inequality,

$$\frac{d}{dt}\|\theta(t) - \theta^*\|_p \leq \|f_1(\theta(t)) - f_1(\theta^*)\|_p - \|f_2(\theta(t)) - f_2(\theta^*)\|_p.$$

By choosing $p = \infty$,

$$\frac{d}{dt}\|\theta(t) - \theta^*\|_\infty \leq -\gamma \|\theta(t) - \theta^*\|_\infty. \tag{7.31}$$

The solution of the ODE (7.31) is

$$\|\theta(t) - \theta^*\|_\infty \leq e^{-\gamma t}\|\theta(0) - \theta^*\|_\infty. \tag{7.32}$$

The solution (7.32) means that the equilibrium point θ^* of (7.27) is globally asymptotically stable.

Because

$$f_1(\theta^*) + f_2(\theta^*) = 0 \Rightarrow f_1(\theta^*) = -f_2(\theta^*)$$

at the equilibrium point, the equilibrium point θ^* is obtained by minimizing the $\hat{Q}$ function,

$$\theta^* = \Phi^\dagger(t) \left(r(t+T) + (1 - \gamma T) \min_{f_d} \Phi^{\mathsf{T}}(t)\theta^* \right), \tag{7.33}$$

where $\Phi^\dagger(\cdot) = \left(\Phi\Phi^{\mathsf{T}} \right)^{-1}\Phi$ is de Moore-Penrose pseudoinverse.

Rewrite (7.27) as

$$\dot{\theta}(t) = \alpha(t)Y(\theta(t), O(t)) = f(\theta(t)), \tag{7.34}$$

where $O(t)$ is a function of the components of all data that depends on the pair $[e_r(t), f_d(t)]$. Since $Y(\cdot)$ is made up of RBFs, $Y(\theta(t), O(t))$ is bounded as

$$\|Y(\theta(t), O(t))\|_\infty \leq \chi(1 + \|\theta(t)\|_\infty), \tag{7.35}$$

where $\chi > 0$. The bound (7.35) means that Y does not depend on the state-action pair but only on the parameter vector $\theta(t)$. When the conditions (7.26), (7.32), and (7.35) are satisfied, there exists a positive definite function $U(\theta(t)) \in C^2$ with bounded second derivatives that satisfies [18]

$$\frac{dU(\theta(t))}{dt} = \frac{\partial U(\theta(t))}{\partial \theta(t)} f(\theta(t)) \leq 0, \quad U(\theta(t)) = 0, \text{ iff } \theta(t) = \theta^*. \tag{7.36}$$

Notice that since $\alpha(t) \in (0, 1]$, (7.26) requires that all state-action pairs be visited infinitely often, which is analogous to the persistently exciting (PE) condition. Note that in (7.36) for any $\frac{\partial U(t)(\theta(t))}{\partial \theta(t)} > 0$, its first derivative tends to zero sufficiently rapidly. From Theorem 17 of [18], $\theta(t)$ converges to a near-optimal value θ^*, and the continuous time reinforcement learning (7.24) converges to $\hat{Q}^*(e_r, f_d; \theta^*)$. ∎

Reinforcement learning has one important property called delayed reward, i.e., the reward received at time t cannot decide whether the policy $h(e_r)$ is good or not. We have to wait for the final results. The parameter update rule (7.27) is a one-step law. In order to use previous history of states and actions, we use exponential traces to modify the temporal profile of the Q-function as

$$Q(e_r(t), f_d(t)) = \begin{cases} e^{-\gamma(T-t)} & \text{if } t \leq T, \\ 0 & \text{if } t > T. \end{cases} \tag{7.37}$$

The profile (7.37) represents the impulse reward at time $t = T$, which takes into account the previous visited states. Since the Q-function is linear with respect to the reward function $r(t + T)$, the correction for the error $\delta(T)$ is

$$W(t) = \begin{cases} \delta(T) e^{-\gamma(T-t)} & \text{if } t \leq T, \\ 0 & \text{if } t > T. \end{cases} \tag{7.38}$$

The parameter update law is modified as

$$\begin{aligned} \dot{\theta}(t) &= \alpha(t) \int_{-\infty}^{T} W(t) \frac{1}{T} \frac{\partial \hat{Q}(e_r(t), f_d(t); \theta(t))}{\partial \theta} dt \\ &= \alpha(t) \frac{\delta(T)}{T} \int_{-\infty}^{T} e^{-\gamma(T-t)} \frac{\partial \hat{Q}(e_r(t), f_d(t); \theta(t))}{\partial \theta(t)} dt. \end{aligned}$$

The lower limit of the integral is $-\infty$ because we look backward to the visited state-action pair. The exponential weighted integral is considered as the eligibility trace of θ,

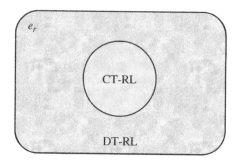

Figure 7.1 Hybrid reinforcement learning.

$$\dot{\theta}(t) = \alpha(t)\delta(t)e(t), \tag{7.39}$$

$$\dot{e}(t) = -\lambda e(t) + \frac{1}{T}\frac{\partial \hat{Q}(e_r(t), f_d(t); \theta(t))}{\partial \theta(t)}. \tag{7.40}$$

The continuous-time reinforcement learning has decay factor λ. It is called $Q(\lambda)$-learning (7.39), similar to the continuous-time Q-learning (7.24). The convergence of (7.39)-(7.40) is similar to Theorem 7.2 where the eligibility trace presents input-state stability [19].

Hybrid Reinforcement Learning

We have discussed how to find the optimal desired force $f_d^*(t)$ in discrete time (DT) and continuous time (CT). When the problem is in the continuous state-action space, we need to use the CT reinforcement learning (RL). It requires more computational effort, since CT-RL has to visit all combinations of state-action pairs.

In order to avoid a large computational effort, we may use hybrid RL, which is shown in Figure 7.1. The diagram shows that the CT-RL is used only in a ball of radius μ, where the position error e_r satisfies the condition $|x - x_e| \leq \mu$; otherwise, we use DT-RL. The position error value depends on how the admittance model gains are chosen according to the environment parameters [20]. When the environment is stiffer than the robot end effector then $x \approx x_e$. On the other hand, when the end effector is stiffer than the environment, then $x \approx x_d$.

The hybrid RL is used to avoid a large exploration phase in states. The PID control guarantees position tracking, the position error has large values at the transient time, and there is no contact force.

7.4 Experiments

Example 7.1 *RL with K-means Clustering*
We use the 2-DOF pan and tilt robot shown in Figure 6.4 to show how the reinforcement learning with K-means clustering achieves the position/force control task and compare it with the LQR optimal solution.

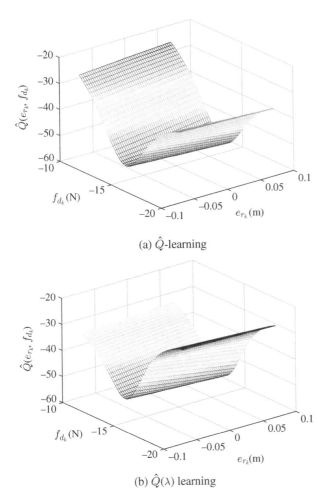

(a) $\hat{Q}$-learning

(b) $\hat{Q}(\lambda)$ learning

Figure 7.2 Learning curves.

The environment is located at the same position as in the experiment in Section 6.4. We used the same learning parameters of Table 6.1 and the control gains of Table 6.2. We use the K -means clustering algorithm to partition the data. The number of clusters is $K = 10$. The width for the RBF is $1/2\sigma^2$, and the standard deviation is $\sigma = 0.1$. For each input, we use 10 RBF, i.e., one RBF has 10 nodes for the desired force. There are 100 hidden nodes. Since we are using the same robot and environment, the LQR solution is (6.47).

The control results are given in Figure 7.2 and Figure 7.3. Since the state-action space is big, the classical Q-learning in discrete-time cannot handle the problem. We use K-means to partition the large space and NRBF to approximate

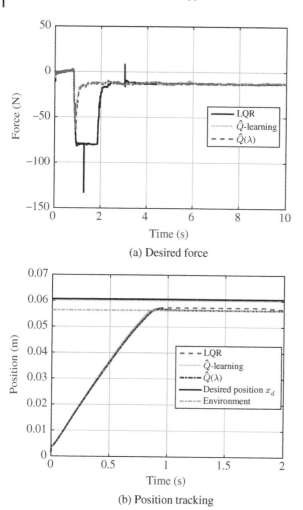

Figure 7.3 Control results.

(a) Desired force

(b) Position tracking

the Q-function. The Q-value approximation algorithms converge to the near-optimal solutions. The use of eligibility trace factor smooths the Q-function learning curve and accelerates its convergence as shown in Figure 7.2. The admittance control achieves both force and position tracking (see Figure 7.3).

Example 7.2 *Continuous-Time Approximation*
We verify the continuous-time and the hybrid reinforcement learning using the 2-DOF pan and tilt robot with a 6-DOF force/torque (F/T) sensor (see Figure 6.4).

We first compare our hybrid reinforcement learning method with theoretical results (optimal solution with the algebraic Riccati equation). The dynamic environment has the parameters of (6.1), but from "C1: $C_e = 200$ Ns/m, $K_e = 1000$ N/m" to "C2: $C_e = 500$ Ns/m, $K_e = 2000$ N/m." The weights of the cost function are $S = 1$ and $R = 0.1$. The gain L in (7.15) is changed, from "C1: $L = 0.005$" to "C2: $L = 0.0025$". The desired Cartesian position is changed from $x_d = [0.0495, 0, 0.1446]^T$ to $x_d = [0.0606, 0, 0.1301]^T$.

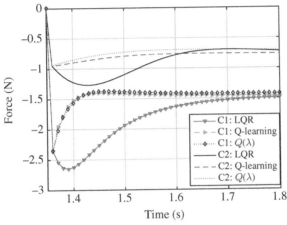

(a) Desired forces learning in different environments

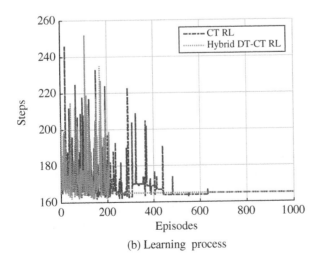

(b) Learning process

Figure 7.4 Hybrid RL in unknown environments.

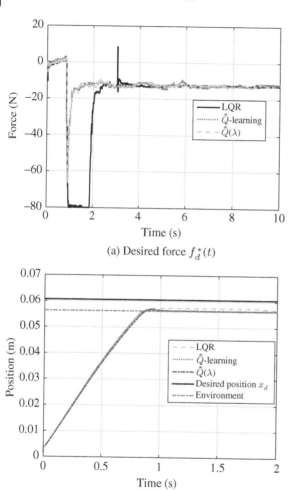

Figure 7.5 Comparisons of different methods.

(a) Desired force $f_d^*(t)$

(b) Position tracking

The learning parameters for both DT-RL and CT-RL are the same as in Table 6.1. The learning has 1,000 episodes with 1,000 steps per episode. Before training, we use the K-means clustering algorithm to partition the data. The number of clusters are $K = 10$, and we use random centers in order to generate a family of approximators. For the neural approximator, the width of RBF is $1/2\sigma^2$, and the standard deviation is $\sigma = 0.1$. For each input, we use 10 RBFs, and each RBF has 10 nodes.

At the training phase, the hybrid RL is used to find the best parameters θ that approximate the value function Q. The desired forces in different environments are obtained by the hybrid RL. The LQR controller solution are shown in Figure 7.4(a). We can see both reinforcement learning algorithms converge.

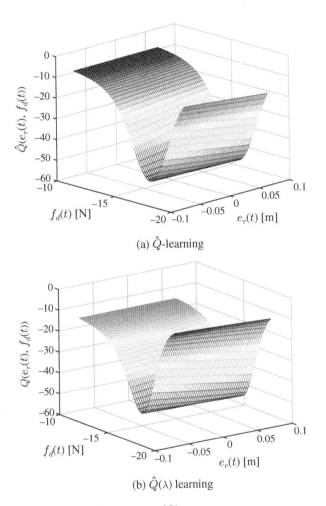

(a) $\hat{Q}$-learning

(b) $\hat{Q}(\lambda)$ learning

Figure 7.6 Learning process of RL.

The continuous-time RL has near-optimal solutions close to the LQR. The use of eligibility trace factor λ smooths the surface. The Q(λ)-learning is better than the normal Q-learning.

We also compare the learning time of the CT-RL and the hybrid RL. The results are shown in Figure 7.4(b). It can be seen that our approach accelerates the learning phase.

Finally we show the experimental setup. The PID controller and the admittance controller gains are given in Table 6.2. The LQR solution gain is computed by (7.15). The control gain is $L = 0.0018$. The control results are given in Figure 7.5 and Figure 7.6.

Discussion

The hybrid RL achieves sub-optimal performance without knowledge of the environment dynamics. The state-action space is continuous. We use the K-means algorithm to partition the input space and RBF neural networks to approximate the Q-function. The hybrid algorithm helps to accelerate convergence to the near-optimal desired force. The use of eligibility traces smooths the Q-function learning curve (see Figure 7.6).

The learning time can be reduced with the hybrid RL by using only the CT-RL in a small subset of the position error. The sub-optimal desired force is found, while the DT-RL deals with states that do not give useful information for the control task and exploits the current knowledge of the CT-RL.

7.5 Conclusions

In this chapter we give the approaches to deal with the high dimension space issue in reinforcement learning for the position/force control task. The algorithms are developed in both discrete and continuous time. Also an hybrid reinforcement learning is used to join the qualities of discrete-time and continuous-time reinforcement learning approaches. This new approach finds the sub-optimal control policy with less learning time.

Simulations and experiments are given to verify the reinforcement learning approximation in a position/force control task. Both position and force tracking are guaranteed.

References

1 M. Sugiyama, H. Hachiya, C. Towell, and S. Vijayakumar, "Value function approximation on non-linear manifolds for robot motor control," *Proceedings of the IEEE International Conference on Robotics and Automation (ICRA'07)*, 2007.

2 L. Buşoniu, R. Babûska, B. De Schutter, and D. Ernst, *Reinforcement learning and dynamic programming using function approximators*. CRC Press, Automation and Control Engineering Series, 2010.

3 I. Grondman, M. Vaandrager, L. Buşoniu, R. Babûska, and E. Schuitema, "Efficient model learning methods for actor-critic control," *IEEE Transactions on Systems, man, and cybernetics. Part B: Cybernetics*, vol. 42, no. 3, 2012a.

4 A. Jain, "Data clustering: 50 years beyond k-means,"*Pattern Recognition Letters*, 2010.

5 W. Barbakh and C. Fyfe, "Clustering with reinforcement learning," *In:* Yin H., Tino P., Corchado E., Byrne W., Yao X. *(eds.) Intelligent Data Engineering and Automated Learning*, vol. 4881, 2007. Berlin, Heidelberg.

6 X. Guo and Y. Zhai, "K-means clustering based reinforcement learning algorithm for automatic control in robots." ISSN:1473-804x online, 1473-8031.

7 S. Bose and M. Huber, "Semi-unsupervised clustering using reinforcement learning," *Proceedings of the Twenty-Ninth International Florida Artificial Intelligence Research Society Conference*, no. 150–153, 2016.

8 N. Tziortziotis and K. Blekas, "Model-based reinforcement learning using on-line clustering," *IEEE 24th International Conference on Tools with Artificial Intelligence (ICTAI)*, 2012.

9 A. Likas, "A reinforcement learning approach to online clustering," *Neural Computation*, vol. 11, pp. 1915–1932, 1999.

10 O. Abul, F. Polat, and R. Alhajj, "Multi-agent reinforcement learning using function approximation," *IEEE transactions on Systems, Man and Cybernetics Part C: Applications and reviews*, 2000.

11 D. Luviano and W. Yu, "Continuous-time path planning for multi-agents with fuzzy reinforcement learning," *Journal of Intelligent & Fuzzy Systems*, vol. 33, pp. 491–501, 2017.

12 K. Doya, "Reinforcement learning in continuous time and space," *Neural Computation*, vol. 12, no. 1, pp. 219–245, 2000.

13 J. Y. Lee, J. B. Park, and Y. H. Choi, "Integral reinforcement learning for continuous-time input-affine nonlinear systems with simultaneous invariant explorations," *IEEE Transactions on Neural Networks and Learning Systems*, vol. 26, no. 10, pp. 2301–2310, 2014.

14 H. M. Palanisamy, F. L. Lewis, and M. Aurangzeb, "Continuous-time q-learning for infinite-horizon discounted cost linear quadratic regulator problems," *IEEE Transactions on Cybernetics*, vol. 45, no. 2, pp. 165–176, 2015.

15 B. Kiumarsi, K. G. Vamvoudakis, H. Modares, and F. L. Lewis, "Optimal and autonomous control using reinforcement learning: A survey," *IEEE Transactions on Neural Networks and Learning Systems*, vol. 29, no. 6, 2018.

16 R. Sutton and B. A, *Reinforcement Learning: An Introduction*. Cambridge, MA: MIT Press, 1998.

17 F. Lewis, D. Vrable, and K. Vamvoudakis, "Reinforcement learning and feedback control using natural decision methods to desgin optimal adaptive controllers," *IEEE Control Systems Magazine*, 2012.

18 A. Benveniste, M. Mativier, and P. Priouret, *Adaptive Algorithms and Stochastic Approximations*, vol. 22. Springer-Verlag: Berlin, 1990.

19 H. K. Khalil, *Nonlinear Systems*. Upper Saddle River, New Jersey: Prentice Hall, 3rd ed., 2002.

20 A. Perrusquía, W. Yu, and A. Soria, "Position/force control of robots manipulators using reinforcement learning," *Industrial Robot: the international journal of robotics research and application*, vol. 46, no. 2, pp. 267–280, 2019.

8

Robot Control in Worst-Case Uncertainty Using Reinforcement Learning

8.1 Introduction

The objective of robust control is to achieve robust performance in presence of disturbances. Most robust controllers are inspired by optimal control theory, such as $\mathcal{H}_2$ control [1], which minimizes a certain cost function to find an optimal controller. The most popular $\mathcal{H}_2$ controller is the linear quadratic regulator [2]. However, it does not work well in presence of disturbances. The control algorithm $\mathcal{H}_\infty$ can find a robust controller when the system has disturbances, although its performance is poor [3]. The combination of controllers $\mathcal{H}_2$ and $\mathcal{H}_\infty$, called control $\mathcal{H}_2/\mathcal{H}_\infty$, has both advantages and has optimal performance with bounded disturbance [4]. Nevertheless, the controller design needs a complete knowledge of the system dynamics [5]. These controllers are model-based.

Model-free controllers, such as the PID control [6, 7], SMC [8] and the neural control [9–11], do not require dynamic knowledge of the system, as it was shown in Chapter 4. However, parameter tuning and some prior knowledge of the disturbances prevent these model-free controllers to perform optimally.

Recent results show that reinforcement learning methods can learn controllers $\mathcal{H}_2$ and $\mathcal{H}_\infty$ without system dynamics [12]. The main objective of RL for $\mathcal{H}_2$ and $\mathcal{H}_\infty$ is to minimize the total cumulative reward. For a robust controller, the reward is designed in the sense of control problems $\mathcal{H}_2$ and $\mathcal{H}_\infty$. This robust reward can be static or dynamic. When the state-action space is small, the approximate optimal solution can be obtained using any RL method. When the space is large or continuous, RL does not converge to the optimal solution [13, 14], which is also called the curse of dimensionality.

Human-Robot Interaction Control Using Reinforcement Learning, First Edition. Wen Yu and Adolfo Perrusquía.
© 2022 The Institute of Electrical and Electronics Engineers, Inc. Published 2022 by John Wiley & Sons, Inc.

There are two problems when RL is applied in the design of robust controllers:

1. RL overestimates the value functions using the minimum operator in the update rule; therefore, the control policy is affected. The double estimator algorithm [15] can be used to solve the overestimation problem. The double Q-learning algorithm [16] uses two independent estimators to calculate the value function and the policy, but there is the possibility of underestimation in some action values [16]. This algorithm can not be applied in a robust controller because the state-action space under the worst-case uncertainty[1] is large.

2. When the state-action space is very large, as learning the worst case, RL becomes impossible [17]. Some approximations, such as the radial basis functions (RBF) neural networks (NN), which we already have discussed in the last chapter and the k-nearest neighbors (kNN) [18], are used to solve the second problem. The kNN method is a nonparametric estimator that does not require any information from the unknown nonlinear system. It is suitable for a large state-action space [18], but it also has the overestimation problem in the action value function estimation.

The RBF-NN approximation requires a lot of prior information for the unknown nonlinear system [19] because they are localizable functions. When the above information is not available, the actor-critic (AC) methods are used [20, 21], which use two estimators, such as neural networks, to complete the policy search work [12]. However, the control depends on the accuracy of the neural estimators [13, 22]. Most robust controllers using reinforcement learning are based on the actor-critic algorithms and Q-learning [23–26].

Usually the robust controller is designed in the form of a zero-sum differential game [23], which requires partial knowledge of the system dynamics to obtain the optimal and robust control policy [23, 24]. To avoid any knowledge of the dynamics, an off-policy method is used such as Q-learning to find the optimal control input and worst-case disturbance [25, 26]. The neural networks are used as the approximators to deal with a large state-action space, but they have the overestimation problem, and its accuracy lies in the neural network design.

In this chapter, we modify the RL methods to overcome the overestimation and the dimension problems for the robust control problem under the worst-case uncertainty. The new robust controller does not need dynamic knowledge and avoids the initialization of a new learning phase.

---

1 The worst-case uncertainty is the worst perturbation that the system can present.

8.2 Robust Control Using Discrete-Time Reinforcement Learning

We use the subscript t to denote the time index instead of k. Consider the following nonlinear dynamic system

$$x_{t+1} = f(x_t) + g_1(x_t)\omega_t + g_2(x_t)u_t, \tag{8.1}$$

where $f(x_t) \in \mathbb{R}^n, g_1(x_t) \in \mathbb{R}^{n \times \omega}$, and $g_2(x_t) \in \mathbb{R}^{n \times m}$ define the dynamics of the system, $x_t \in \mathbb{R}^n$ is the state, $u_t \in \mathbb{R}^m$ is the control input, and the disturbance is described as $\omega_t \in \mathbb{R}^\omega$.

If $\omega_t = 0$ and the control $u(x_t)$ is an admissible control law that minimizes the following index

$$J_2(x_t, u_t) = \sum_{i=0}^{\infty} \left(x_i^\top S x_i + u_i^\top R u_i \right), \tag{8.2}$$

then the controller $u(x_t)$ is called the solution of $\mathcal{H}_2$ control, where $S \in \mathbb{R}^{n \times n}$ and $R \in \mathbb{R}^{m \times m}$ are the weight matrices of the $\mathcal{H}_2$ problem. Hence, the problem to be solved can be defined as

$$V(x_t) = \min_u \left(\sum_{i=t}^{\infty} \left(x_i^\top S x_i + u_i^\top R u_i \right) \right). \tag{8.3}$$

The value function (8.3) can be defined as the Bellman equation

$$V(x_t) = \sum_{i=t}^{\infty} \left(x_i^\top S x_i + u_i^\top R u_i \right) = r_{t+1} + V(x_{t+1}),$$

where $r_{t+1} = x_t^\top S x_t + u_t^\top R u_t$ is the reward function or utility function. Bellman's optimality principle yields the optimal value function

$$V^*(x_t) = \min_u \left[r_{t+1} + V^*(x_{t+1}) \right].$$

The optimal control is derived as

$$u^*(x_t) = \arg\min_u \left[r_{t+1} + V^*(x_{t+1}) \right] = -\frac{1}{2} R^{-1} g_2^\top(x) \frac{\partial V^*(x_{t+1})}{\partial x_{t+1}}. \tag{8.4}$$

If $\omega_t \neq 0$ and the control $u(x_t)$ is an admissible that minimizes the index

$$J_\infty(x_t, u_t, \omega_t) = \sum_{i=0}^{\infty} \left(x_i^\top S x_i + u_i^\top R u_i - \eta^2 \omega_i^\top \omega_i \right), \tag{8.5}$$

then $u(x_t)$ is called the solution of $\mathcal{H}_\infty$ control. Here η is an attenuation factor. Similarly to the $\mathcal{H}_2$ control case, the value function is defined as

$$V(x_t) = x_t^\top S x_t + u_t^\top R u_t - \eta^2 \omega_t^\top \omega_t + V(x_{t+1}).$$

The optimal value can be obtained by solving a zero-sum differential game as

$$V^*(x_t) = \min_u \max_\omega J_\infty(x_t, u, \omega).$$

To solve the zero-sum game, it is required to solve the Hamilton-Jacobi-Isaacs (HJI) equation

$$V^*(x_t) = x_t^\top S x_t + u_t^{*\top} R u_t^* - \eta^2 \omega_t^{*\top} \omega_t^* + V^*(x_{t+1}).$$

The optimal control with respect to the worst-case uncertainty are

$$u^*(x_t) = -\frac{1}{2}R^{-1}g_2^\top(x)\frac{\partial V^*(x_{t+1})}{\partial x_{t+1}}, \tag{8.6}$$

$$\omega_t^* = \frac{1}{2\eta^2}g_1^\top(x)\frac{\partial V^*(x_{t+1})}{\partial x_{t+1}}. \tag{8.7}$$

where (8.6) is to the robust control solution, (8.7) is the "optimal" worst-case uncertainty, which is the maximum disturbance for which the system can handle the optimal performance.

Generally, $\mathcal{H}_2$ control does not have good performance against perturbations, while $\mathcal{H}_\infty$ control has poor control performances and good robustness. Therefore, $\mathcal{H}_2/\mathcal{H}_\infty$ control [1] is developed, which has both advantages. Here we deal with an optimization problem with constraints:

$$\min_u J_2(x_t, u)$$

subject to: $J_\infty(x_t, u_t, \omega_t) \leq 0.$ \hfill (8.8)

One way to obtain the $\mathcal{H}_2/\mathcal{H}_\infty$ control is using a parametric case [1] as

$$u_\xi^*(x_t) = \xi u(x_t) + (\xi - 1)R^{-1}g_2^\top(x)\frac{\partial J_2^\xi(x_{t+1})}{\partial x_{t+1}}, \tag{8.9}$$

where $\xi \in (-1, 1)$ and $u(x_t)$ is any controller that stabilizes the system (8.1). The parameter ξ helps to obtain a family of stabilizing controllers that are robust and optimal, which minimizes (8.2) subject to (8.5).

The "optimal" worst-case uncertainty ω_t^* is different to the worst-case uncertainty of the system $\overline{\omega}$. The worst disturbance $\overline{\omega}$ is the maximum disturbance that the system can present, i.e., $0 \leq \omega_t^* \leq \overline{\omega}$.

All above controllers ($\mathcal{H}_2$, $\mathcal{H}_\infty$, and $\mathcal{H}_2/\mathcal{H}_\infty$) need knowledge of the system dynamics. When the dynamics of the nonlinear system (8.1) are unknown, we can use reinforcement learning to obtain a solution of the robust control problem instead of using the value function $V(x_t)$. We use the action-value function $Q(x_t, u_t)$ to consider the actual control action performance. We want to design a robust controller that is optimal or near optimal with respect to

$$Q(x_t, u_t) = \sum_{i=t}^{\infty} \gamma^{i-t} \left(x_i^T S x_i + u_i^T R u_i \right) = \sum_{i=t}^{\infty} \gamma^{i-t} r_{i+1} \tag{8.10}$$

$$\text{subject to: } \|u\| < \bar{u}, \quad \|x\| < \bar{x}, \quad \|\omega\| \leq \bar{\omega}, \tag{8.11}$$

where $\bar{u}, \bar{x}$, and $\bar{\omega}$ are known upper bounds of the variables.

The discount factor serves to weight (when $\gamma < 1$) the past rewards, which is very useful when the system is unchanged in future time steps. In the worst-case uncertainty, it is preferable to use the discount factor $\gamma = 1$ to take into account possible changes in the system dynamics. The robust reward (8.10) can be either static or dynamic. The typical reward is the quadratic form and is solved using convex optimization methods [27].

The robust and near-optimal value function is defined as the best Q-function that can be obtained by any control policy u_t

$$Q^*(x_t, u_t) = \min_u Q(x_t, u).$$

The Bellman equations of Q and Q^* are given as follows:

$$Q(x_t, u_t) = r_{t+1} + \gamma Q(x_{t+1}, u(x_{t+1})),$$
$$Q^*(x_t, u_t) = r_{t+1} + \gamma \min_u Q^*(x_{t+1}, u). \tag{8.12}$$

By using the cost-to-go value function and the Bellman equations, we can estimate the value function Q in (8.10). Q-learning (see Appendix B) can be estimated by using the following update rule:

$$Q_{t+1}(x_t, u_t) = Q_t(x_t, u_t) + \alpha_t \left(q_t - Q_t(x_t, u_t) \right), \tag{8.13}$$

where Q_t is the state-action value function of the state x_t and action u_t, and q_t is the learning target. For Q-learning, the learning target is

$$q_t = r_{t+1} + \gamma \min_u Q_t(x_{t+1}, u),$$

which is equivalent to the right-hand term of (8.12).

The main goal of reinforcement learning (8.13) methods is to minimize the total cumulative reward to arrive at $\mathcal{H}_2/\mathcal{H}_\infty$ as in (8.10). Since the control and the disturbances are bounded as shown in (8.11), the control policy u_t will not be the optimal control policy for (8.10) and the Q-value cannot arrive the optimal function Q^*. By means of (8.12), the control policy is near to the optimal control policy, and it is stable with respect to the worst-case uncertainties. Thus, it is robust and sub-optimal under the worst-case uncertainty.

8.3 Double Q-Learning with k-Nearest Neighbors

The worst-case learning needs a large state-action space. An on-policy learning rule, such as Sarsa [15] (see Appendix B), uses greedy exploration techniques to explore the whole space. An off-policy learning rule, such as Q-learning, minimizes the action-value with the estimated action values. It also uses the exploration techniques. The action value minimization and the exploration techniques may lead to overestimation of action values and policies and may not converge to the near-optimal solution because the state-action space is huge in the worst case.

The update rule (8.13) can be derived from the expected value,

$$\mu = \sum_{i=1}^{n} q_i p(q_i),$$

where $q_1, q_2, ..., q_n$ are discrete possible random variables of μ, For two possible values (a, b), the above expression becomes

$$\mu = (1 - \alpha)a + b\alpha,$$

where α is the probability of the second value b. If we take a as a previously stored value μ and b as a new observation q then

$$\mu = \mu + \alpha\,(q - \mu),$$

which is equivalent to the TD-learning update rule (8.13) by replacing μ with $Q(x, u)$. We use the k-nearest neighbors (kNN) rule to reduce the computational effort of reinforcement learning methods in large state-action spaces by using only k states for the state-action value function learning.

The distance metric used to identify the nearest neighbor is the Euclidean metric,

$$d = \sqrt{\sum_{i=1}^{n}(x_i - x)^2},$$

where $x_1, x_2, ..., x_n$ are the training examples. The label of the test example is determined by the majority vote of its k nearest neighbors. Each neighbor has an associated action-value function predictor $Q(i, u)$ and a weight w_i,

$$w_i = \frac{1}{1 + d_i^2}, \forall i \in 1, ..., k$$

For each time step, there are k active classifiers, whose weights are the inverses of the distances from the current state x to its neighbors [18].

The kNN set is defined by the k states that minimize the distance metric from the current state. After the kNN set is obtained in the state space, the probability distribution $p(kNN)$ over kNN is calculated. These probabilities are implied by the current weight vector $\{w\}$. It is normalized in order to express the probability distribution,

$$p(i) = \frac{w_i}{\sum w_i} \quad \forall i \in \text{kNN}. \tag{8.14}$$

For the control action, the expectation for each action should include all prediction process. This process involves many different predictions. We use another kNN algorithm to get one element from each kNN set. The expected value of the learning target is

$$\langle Q(kNN, u) \rangle = \sum_i^{i=kNN} Q(i, u)p(i), \tag{8.15}$$

where $\langle Q(kNN, u) \rangle$ is the value of $Q_t(i, u)$, and $p(i)$ is the probability $Pr\left[Q_t(i, u)|x_t\right]$. The control action is derived by the greedy policy

$$u^* = \arg\min_u Q(kNN, u). \tag{8.16}$$

The Q-learning algorithm (8.13) relies on past and present observations, actions, and the received reward $(x_t, u_t, x_{t+1}, r_{t+1})$. The cumulative reward is obtained by the action-value of the current state as in (8.15). The expected value is calculated as

$$\left\langle Q(kNN_{t+1}, u) \right\rangle = \sum_i^{i=kNN} Q(i, u)p'(i), \tag{8.17}$$

where kNN_{t+1} is the set of the k-nearest neighbors of x_{t+1}, kNN_t is the k-nearest neighbors of x_t, and $p'(i)$ are the probabilities of each neighbor in the kNN_{t+1} set. The update rule for the expected action-value function is

$$Q_{t+1}(i, u_t) = Q_t(i, u_t) + \alpha_t \delta_t \quad \forall i \in kNN_t, \tag{8.18}$$

where δ_t is calculated as

$$\delta_t = r_{t+1} + \gamma \min_u Q_t(kNN_{t+1}, u) - Q_t(kNN_t, u_t). \tag{8.19}$$

The above reinforcement learning method has the overestimation problem because the learning target uses the minimum operator at the update rule. This means that the minimum overestimated action-value functions are used implicitly. It leads to biased estimates. We use the double Q-learning method to overcome this problem. The double Q-learning method is based on the double estimator technique that divides the sample set

$$D = \cup_i^N D_i$$

into two disjoint subsets, D^A and D^B. The unbiased estimators for the two sets are $\mu^A = \{\mu_1^A, ..., \mu_N^A\}$ and $\mu^B = \{\mu_1^B, ..., \mu_N^B\}$. Here

$$E\{\mu_i^A\} = E\{\mu_i^B\} = E\{Q_i\},$$

for all i. The two estimators learn the true value of $E\{Q_i\}$ independently.

Here μ^A is used to determine the minimizing action

$$u^* = \arg\min_u \mu^A(u).$$

μ^B estimates its value,

$$\mu^B(u^*) = \mu^B(\arg\min_u \mu^A(u)).$$

It is unbiased, i.e., $E\{\mu^B(u^*)\} = Q(u^*)$. When the variables are i.i.d., the double estimator is unbiased because all expected values are equal and $\Pr(u^* \in \arg\min{}_i E\{Q_i\}) = 1$.

In terms of the kNN approximator we have two separate Q functions, Q^A and Q^B, and each one is approximated by the kNN set. The two estimators reduce susceptibility to random variation in r_{t+1} and stabilize the action values. Additionally, Q^A and Q^B are switched with probability 0.5. So each estimator is only updated using half experiences.

Q^A is updated

$$u_t^* = \arg\min_u Q_t^A(kNN_{t+1}, u). \tag{8.20}$$

Q^B is updated

$$v_t^* = \arg\min_u Q_t^B(kNN_{t+1}, u). \tag{8.21}$$

The kNN-TD learning for large state (LS) and discrete action (DA) with the double estimators is

$$\begin{aligned}
\delta_t^{BA} &= r_{t+1} + \gamma Q^B(kNN_{t+1}, u_t^*) - Q^A(kNN_t, u_t), \\
\delta_t^{AB} &= r_{t+1} + \gamma Q^A(kNN_{t+1}, v_t^*) - Q^B(kNN_t, u_t), \\
Q_{t+1}^A(i, u_t) &= Q_t^A(i, u_t). + \alpha_t \delta_t^{BA}, \\
Q_{t+1}^B(i, u_t) &= Q_t^B(i, u_t). + \alpha_t \delta_t^{AB} \ \forall i \in kNN_t.
\end{aligned} \tag{8.22}$$

For the large state (LS) and large action (LA) space, the action is independent of the value $Q(kNN_{t+1}, u)$. Given an action list $\mathcal{U}$, the best action in each kNN set is an optimal action list U^* given by

$$I = \arg\min_u Q(i, u), \quad U^* = \mathcal{U}[I], \ \forall i \in kNN \text{ and } \forall u \in \mathcal{U}, \tag{8.23}$$

where $\mathcal{U}$ are the actions where the values of $Q(i, u)$ are minimal. The expected optimal action is

$$\langle u \rangle = \sum_{i=1}^{kNN} U^*(i) p(i), \ \forall i \in kNN,$$

where $p(i)$ is the conditional probability $\Pr\{u = U^*(i)|x\}$, when u takes the value $U^*(i)$ given the state x.

The TD-error as in (8.19) is modified as

$$\delta_t = r_{t+1} + \gamma Q(kNN_{t+1}, I_{t+1}) - Q(kNN_t, I_t). \tag{8.24}$$

The action values are calculated by

$$\langle Q(kNN_t, I_t) \rangle = \sum_{i=1}^{kNN_t} \min_{\mathcal{J}} Q(i, \mathcal{J}) p(i)$$

$$\langle Q(kNN_{t+1}, I_{t+1}) \rangle = \sum_{i=1}^{kNN_{t+1}} \min_{\mathcal{J}} Q(i, \mathcal{J}) p(i),$$

where $\mathcal{J}$ is the list $1...n = |\mathcal{U}|$. Using the double estimator algorithm, the action values are given by

$$Q^A(kNN_t, I_t) = Q(kNN_t, u) p(kNN_t),$$

$$Q^B(kNN_{t+1}, I_{t+1}) = Q(kNN_{t+1}, u) p(kNN_{t+1}). \tag{8.25}$$

Unlike the LS-DA case, LS-LA does not require that the updates are switched with probability 0.5; instead it uses a serial learning. The minimizing actions for the estimators are

$$u_t^A = \langle u^A \rangle = U^*(kNN_t) p(kNN_t),$$

$$u_t^B = \langle u^B \rangle = U^*(kNN_{t+1}) p(kNN_{t+1}). \tag{8.26}$$

The minimizing actions are obtained by the best recommended action U^* given the kNN_t and kNN_{t+1} sets. The update rule of LS-LA is

$$Q_{t+1}(i, u_t) = Q_t(i, u_t) + \alpha_t \delta_t \ \forall i \in kNN_t,$$

$$\delta_t = r_{t+1} + \gamma Q^B(kNN_{t+1}, u_t^B) - Q^A(kNN_t, u_t^A). \tag{8.27}$$

Here the estimators Q^A and Q^B are completely independent. However, at each final step, the new value of Q^A is Q^B, i.e., $Q^A \leftarrow Q^B$. The algorithm keeps going until x is terminal.

The following algorithm gives the detail steps of the LS-DA (large states and discrete actions) and LS-LA (large states and large actions) methods, where cl are the number of classifiers.

Convergence Property of the Modified Reinforcement Learning

In this section, we prove that the reinforcement learning with the kNN and double Q-learning modifications guarantees the robust controller under worst-case uncertainty convergence with a near-optimal value.

Algorithm 8.1 Reinforcement learning with kNN and double Q-learning modification: LS-DA case

---

1: **Input:** Worst perturbation $\bar{\omega}$
2: Initialize $Q_0^A(cl, u)$ and $Q_0^B(cl, u)$, cl arbitrarily
3: **repeat** {for each episode}
4:     From x_0 we obtain the kNN_0 set and its probabilities $p(kNN_0)$ using (8.14)
5:     Calculate Q_0^A and Q_0^B based on (8.15) using Q^A and Q^B.
6:       Then choose u_0 from x_0 according to $Q_0^A(\cdot)$ and $Q_0^B(\cdot)$
7:     **repeat** {for each step of episode $t = 0, 1, ...$}
8:       Take action u_t and observe r_{t+1}, x_{t+1}
9:       Calculate the kNN_{t+1} set and its probabilities $p(kNN_{t+1})$ using (8.14)
10:      Calculate Q_0^A, Q_0^B Q^A, and Q^B from (8.17).
11:      Use (8.20) and (8.22) to update Q_t^A. Or use (8.21) and (8.22) to update Q_t^B.
12:      Then $x_t \leftarrow x_{t+1}, kNN_t \leftarrow kNN_{t+1}, u_t \leftarrow u_{t+1}$.
13:     **until** x_t is terminal
14: **until** learning ends

Algorithm 8.2 Reinforcement learning with kNN and double Q-learning modification: LS-LA case

---

1: **Input:** Worst perturbation $\bar{\omega}$
2: Initialize cl, $\mathcal{V}$, and $Q_0(cl, \mathcal{J})$ arbitrarily
3: **repeat** {for each episode}
4:     From x_0 we obtain the kNN_0 set and its probabilities $p(kNN_0)$ using (8.14)
5:     Calculate action index I_0 using (8.23) and control action u_0^A by (8.26). Then obtain Q_0^A using (8.25).
6:     **repeat** {for each step of episode $t = 0, 1, ...$}
7:     Take action u_t and observe r_{t+1}, x_{t+1}
8:     Calculate the kNN_{t+1} set and its probabilities $p(kNN_{t+1})$ using (8.14)
9:     Calculate the action index I_{t+1} using (8.25) and control action u_t^B by (8.27).
10:     Calculate Q_t^B using (8.25). Update Q_t using (8.27).
11:     Then $x_t \leftarrow x_{t+1}, kNN_t \leftarrow kNN_{t+1}, u_t^A \leftarrow u_{t+1}^B, I_t \leftarrow I_{t+1}, Q_t^A \leftarrow Q_t^B$.
12:     **until** x_t is terminal
13: **until** learning ends

The following lemma gives the convergence property of kNN.

Lemma 8.1 *[28] Let x, $x_1, x_2, ..., x_n$ be independent identically distributed random variables, taking values in a separable metric space X, $y_k(x)$ is the kth nearest neighbor of x from the set $\{x_1, x_2, ..., x_n\}$, $N_k(x) = \{y_1(x), ..., y_k(x)\}$ is the set of the k-nearest neighbors. Then for a given k, when the sample size $n \to \infty$, $\|y_k(x) - x\| \to 0$ with probability one.*

This convergence property assures that the kth nearest neighbor to x_i converges to x with probability one, when the data number n goes to infinity. A larger k has better performance. When $k \to \infty$, the approximation error rate approaches the Bayesian error [29]. The convergence proof of this lemma is an extension of the proof the double Q-learning algorithm[16].

Theorem 8.1 *Assume the following conditions are fulfilled: (1) $\gamma \in (0, 1]$, (2) $\alpha_t(x, u) \in (0, 1]$, $\sum_t \alpha_t(x, u) = \infty$, $\sum_t \left(\alpha_t(x, u)\right)^2 < \infty$ w.p. 1 and $\forall(x_t, u_t) \neq (x, u)$: $\alpha_t(x, u) = 0$, and (3) $var\{R_{xu}^{x'}\} < \infty$. Then, given an ergodic MDP, both Q^A and Q^B (LS-DA case) or Q (LS-LA case) will converge on the limit to the near-optimal and robust value function Q^* as given in the Bellman optimal equation (8.12) with probability one if the reward is designed as (8.10) and ensures that each state-action pair is visited an infinite number of times.*

Proof: Let's consider the LS-DA case and that the reward satisfies (8.10). The update rules are symmetric, so it is suffices to show the convergence of one of them. Using the conditional probabilities $\Pr(Q_t^A(kNN, u) = Q_t^A(i, u)|x_t)$ and $\Pr(Q_t^B(kNN, u) = Q_t^B(i, u)|x_t)$ with Lemma 8.1, we obtain that $Q^A(kNN, u) = Q^A(x, u)$ and $Q^B(kNN, u) = Q^B(x, u)$. Applying Lemma B.1 with $Z = X \times U$, $\Delta_t = Q_t^A - Q^*$, $\zeta = \alpha$, $P_t = \{Q_0^A, Q_0^B, x_0, u_0, \alpha_0, r_1, x_1, ..., x_t, u_t\}$, $F_t(x_t, u_t) = r_{t+1} + \gamma Q_t^B(x_{t+1}, u_t^*) - Q^*(x_t, u_t)$, where $u_t^* = \arg\min_u Q^A(x_{t+1}, u)$ and $\kappa = \gamma$. Conditions (2) and (4) hold as a consequence of the conditions in this theorem. Therefore, we only require to prove condition (3) of Lemma B.1. We can write $F_t(x_t, u_t)$ as

$$F_t(x_t, u_t) = G_t(x_t, u_t) + \gamma \left(Q_t^B(x_{t+1}, u_t^*) - Q_t^A(x_{t+1}, u_t^*)\right), \qquad (8.28)$$

where $G_t = r_{t+1} + \gamma Q_t^A(x_{t+1}, u_t^*) - Q^*(x_t, u_t)$ is the value of F_t considering the Q-learning algorithm (B.24). It is known that $E\{G_t|P_t\} \leq \gamma \|\Delta_t\|$, so we have that $c_t = \gamma \left(Q_t^B(x_{t+1}, u_t^*) - Q_t^A(x_{t+1}, u_t^*)\right)$, and it suffices that $\Delta_t^{BA} = Q_t^B - Q_t^A$ converges to zero. The update of Δ_t^{BA} at time t is

$$\Delta_{t+1}^{BA}(x_t, u_t) = \Delta_t^{BA}(x_t, u_t) + \alpha_t(x_t, u_t)F_t^B(x_t, u_t) - \alpha_t(x_t, u_t)F_t^A(x_t, u_t),$$

where $F_t^A(x_t, u_t) = r_{t+1} + \gamma Q_t^B(x_{t+1}, u_t^*) - Q_t^A(x_t, u_t)$ and $F_t^B(x_t, u_t) = r_{t+1} + \gamma Q_t^A(x_{t+1}, v_t^*) - Q_t^B(x_t, u_t)$. Then

$$E\{\Delta_{t+1}^{BA}(x_t, u_t)|P_t\} = \Delta_t^{BA}(x_t, u_t) + E\{\alpha_t F_t^B(x_t, u_t) - \alpha_t F_t^A(x_t, u_t)|P_t\}$$
$$= (1 - \alpha_t)\Delta_t^{BA}(x_t, u_t) + \alpha_t E\{F_t^{BA}(x_t, u_t)|P_t\},$$

where $E\{F_t^{BA}(x_t, u_t)|P_t\} = \gamma E\{Q_t^A(x_{t+1}, v_t^*) - Q_t^B(x_{t+1}, u_t^*)|P_t\}$. Since the update of the action-value function (Q_t^A or Q_t^B) is random, we have the following cases:

i) Assume $E\{Q_t^A(x_{t+1}, v_t^*)|P_t\} \leq E\{Q_t^B(x_{t+1}, u_t^*)|P_t\}$. Then by definition of the action selection v_t^* of Algorithm 8.1 we have that the following inequality is satisfied: $Q_t^A(x_{t+1}, v_t^*) = \min_u Q_t^A(x_{t+1}, u) \leq Q_t^A(x_{t+1}, u_t^*)$; therefore,

$$E\{F_t^{BA}|P_t\} = \gamma E\{Q_t^A(x_{t+1}, v_t^*) - Q_t^B(x_{t+1}, u_t^*)|P_t\}$$
$$\leq \gamma E\{Q_t^A(x_{t+1}, u_t^*) - Q_t^B(x_{t+1}, u_t^*)|P_t\} \leq \gamma \|\Delta_t^{BA}\|.$$

ii) Assume $E\{Q_t^B(x_{t+1}, u_t^*)|P_t\} \leq E\{Q_t^A(x_{t+1}, v_t^*)|P_t\}$. Then by definition of the action selection u_t^* of Algorithm 8.1 we have that the following inequality is satisfied: $Q_t^B(x_{t+1}, u_t^*) = \min_u Q_t^B(x_{t+1}, u) \leq Q_t^B(x_{t+1}, v_t^*)$; therefore,

$$E\{F_t^{BA}|P_t\} = \gamma E\{Q_t^A(x_{t+1}, v_t^*) - Q_t^B(x_{t+1}, u_t^*)|P_t\}$$
$$\leq \gamma E\{Q_t^A(x_{t+1}, v_t^*) - Q_t^B(x_{t+1}, v_t^*)|P_t\} \leq \gamma \|\Delta_t^{BA}\|.$$

One of the presented cases must hold at each time step, and in both cases we obtain the same desired result. Then applying Lemma B.1 yields the convergence of Δ_t^{BA} to zero, and therefore Δ_t converges also to zero, and Q^A and Q^B converge to Q^*, which also is robust. The LS-LA proof is similar to the above proof and the Sarsa convergence proof (B.28) [15]. ∎

In the practical applications, we can simply use constant learning rates, such that the condition $\sum_t \alpha_t(x, u) = \infty$ is satisfied (see the proof of Theorem B.1). One valuable result of the kNN method [28] is that its error is bounded by the Bayes error as

$$P(e_b|x) \leq P(e_{kNN}|x) \leq 2P(e_b|x), \tag{8.29}$$

where $P(e_b|x)$ is the Bayes error rate and $P(e_{kNN}|x)$ is the kNN error rate, which means that the kNN rule error is less than twice the Bayes error,

$$E\left\{\left(\sum_{i=1}^{kNN} Q(i, u)p(i) - Q(x, u)\right)^2\right\} \leq 2E\left\{\min_i \left(\sum_{i=1}^{kNN} Q(i, u)p(i) - Q(x, u)\right)\right\} \tag{8.30}$$

8.4 Robust Control Using Continuous-Time Reinforcement Learning

Consider the continuous time nonlinear system [1]

$$\dot{x}(t) = f(x(t)) + g_1(x(t))u(t) + g_2(x(t))\omega(t), \quad x(t_0) = x(0) \; t \geq t_0, \tag{8.31}$$

where $f(x(t)) \in \mathbb{R}^n, g_1(x(t)) \in \mathbb{R}^{n\times m}, g_2(x(t)) \in \mathbb{R}^{n\times\omega}$ are the dynamics of the non-linear system, $x(t) \in X \subset \mathbb{R}^n$ is the state, $u(t) \in U \subset \mathbb{R}^m$ is the control input, and the disturbance is described as $\omega \in \mathbb{R}^\omega$.

When $\omega = 0$, the $\mathcal{H}_2$ admissible control law $u(t) = h(x(t))$ minimizes

$$J_2(x(0), u) = \int_0^\infty \left(x^T(\tau)Sx(\tau) + u^T(\tau)Ru(\tau) \right) e^{-\gamma(\tau-t)} d\tau. \tag{8.32}$$

where $S \in \mathbb{R}^{n\times n}$ and $R \in \mathbb{R}^{m\times m}$ are positive defined weight matrices for the state and the control input, $\gamma \geq 0$ is a discount factor, and x_0 is the initial state of x_t.

The exponential decay factor $e^{-\gamma t}$ in the cost index helps to guarantee convergence to the optimal solution with the reinforcement learning. The value function for the admissible control is given by the following discounted cost function [12]:

$$V(x(t)) = \int_t^\infty \left(x^T(\tau)Sx(\tau) + u^T(\tau)Ru(\tau) \right) e^{-\gamma(\tau-t)} d\tau, \tag{8.33}$$

Taking the time derivative of (8.33) and using Leibniz's rule, we have the following Bellman equation

$$\dot{V}(x(t)) = \int_t^\infty \frac{\partial}{\partial t} \left[x^T(\tau)Sx(\tau) + u^T(\tau)Ru(\tau) \right] e^{-\gamma(\tau-t)} d\tau$$
$$- x^T(t)Sx(t) - u^T(t)Ru(t). \tag{8.34}$$

The optimal value function is

$$V^*(x(t)) = \min_{\bar{u}[t:\infty)} \int_t^\infty \left(x^T(\tau)Sx(\tau) + u^T(\tau)Ru(\tau) \right) e^{-\gamma(\tau-t)} d\tau, \tag{8.35}$$

where $\bar{u}[t : \infty) := \{u(\tau) : t \leq \tau < \infty\}$. Using Bellman's optimality principle, the Hamilton-Jacobi-Bellman (HJB) equation is

$$\min_{\bar{u}[t:\infty)} \left\{ \dot{V}^*(x(t)) + x^T(t)Sx(t) + u^T(t)Ru(t) \right.$$
$$\left. - \int_t^\infty \frac{\partial}{\partial t} \left(x^T(\tau)Sx(\tau) + u^T(\tau)Ru(\tau) \right) e^{-\gamma(\tau-t)} d\tau \right\} = 0.$$

The Bellman equation $\dot{V}(x(t))$ satisfies

$$\dot{V}^*(x(t)) = \frac{\partial V^*(x(t))}{\partial x(t)} \dot{x}(t). \tag{8.36}$$

Substituting (8.36) into (8.34)

$$\frac{\partial V^*(x(t))}{\partial x(t)} \left(f(x(t)) + g_1(x(t))u(t) \right) = \gamma V^*(x(t)) - x^T(t)Sx(t) - u^T(t)Ru(t). \tag{8.37}$$

The optimal control is obtained by differentiating (8.37) with respect to $u(t)$

$$u^*(t) = -\frac{1}{2}R^{-1}g_1^T(x(t))\frac{\partial V^{*T}(x(t))}{\partial x(t)}. \tag{8.38}$$

If $\omega \neq 0$, the $\mathcal{H}_\infty$ admissible control law $u(x(t))$ minimizes the following cost function:

$$J_\infty(x(0), u, \omega) = \int_0^\infty \left(x^\mathsf{T} S x + u^\mathsf{T} R u - \eta^2 \omega^\mathsf{T} \omega \right) e^{-\gamma(t-\tau)} d\tau, \tag{8.39}$$

where η is the attenuation factor. The value function is

$$V(x(t), u, \omega) = \int_t^\infty \left(x^\mathsf{T} S x + u^\mathsf{T} R u - \eta^2 \omega^\mathsf{T} \omega \right) e^{-\gamma(t-\tau)} d\tau. \tag{8.40}$$

The optimal value can be obtained by solving the following zero-sum differential game

$$V^*(x(t)) = \min_u \max_\omega J_\infty(x(0), u, \omega).$$

In order to obtain the solution, we need the following Hamilton-Jacobi-Isaacs (HJI) equation:

$$x^\mathsf{T}(t)Sx(t) + u^{*\mathsf{T}}(t)Ru^*(t) - \eta^2 \omega^{*\mathsf{T}}(t)\omega^*(t) - \gamma V^*(x(t), u, \omega)$$
$$+ \frac{\partial V^*(x(t), u, \omega)}{\partial x(t)}(f(x(t)) + g_1(x(t))u^*(t) + g_2(x(t))\omega^*(t)) = 0.$$

The solution of the HJI equation is

$$\begin{aligned} u^*(t) &= -\frac{1}{2}R^{-1}g_1^\mathsf{T}(x(t))\frac{\partial V^{*\mathsf{T}}(x(t), u, \omega)}{\partial x(t)}, \\ \omega^*(t) &= \frac{1}{2\eta^2}g_2^\mathsf{T}(x(t))\frac{\partial V^{*\mathsf{T}}(x(t), u, \omega)}{\partial x(t)}. \end{aligned} \tag{8.41}$$

Generally, the $\mathcal{H}_2$ control does not have good performance against perturbations. The $\mathcal{H}_\infty$ control has good robustness performance and bad control performance. So the $\mathcal{H}_2/\mathcal{H}_\infty$ control is developed to have both advantages. The control problem of $\mathcal{H}_2/\mathcal{H}_\infty$ is

$$\begin{aligned} &\min_u J_2(x(0), u) \\ &\text{subject to: } J_\infty(x(0), u, \omega) \leq 0, \end{aligned} \tag{8.42}$$

where J_2 is defined in (8.32), J_∞ is defined in (8.39), the perturbation ω is assumed to be bounded, and the upper bound (the worst-case) $\overline{\omega}$ is known,

This control policy is optimal and robust, which minimizes (8.32) subject to (8.39). This control scheme uses the $\mathcal{H}_2$ control law as the first solution of the problem and then computes J_∞ using (8.39). If $J_\infty < 0$, then the $\mathcal{H}_2$ controller is the solution of the hybrid control; otherwise, the algorithm looks for a subset of controllers that satisfy $J_\infty = 0$. One way to compute the $\mathcal{H}_2/\mathcal{H}_\infty$ controller is using a parametric control law of the form

$$u_\xi^*(t) = \xi u(t) + (\xi - 1)\frac{1}{2}R^{-1}g_1^\mathsf{T}(x(t))\left(\frac{\partial V^{\xi *}(x(t); \xi)}{\partial x(t)} \right)^\mathsf{T}, \tag{8.43}$$

where $\xi \in (-1, 1)$. The parameter ξ helps to obtain a family of stabilizing controllers that are robust and optimal; nevertheless, the controllers (8.38),(8.41), and (8.43) need knowledge of the system dynamics.

Remark 8.1 *The uncertainty $\omega^*(t)$ is the maximum disturbance for which the system can have optimal performance. The worst-case uncertainty of the system $\omega^*(t)$ is the maximum disturbance that the system can present, that is, $0 \le \omega^*(t) \le \bar{\omega}$ [30].*

The continuous-time reward (optimal and robust action-value function) is modified as the $\mathcal{H}_2/\mathcal{H}_\infty$ control problem,

$$Q^*(x(t), \bar{u}[t : t + T)) = \min_{h(x(t))} Q^h(x(t), \bar{u}[t : t + T)) \tag{8.44}$$
$$\text{subject to:} \quad \|u(t)\|, \bar{u}, \ \|x\| < \bar{x}, \ \|\omega\| < \bar{\omega},$$

where $\bar{u}$, $\bar{x}$, and $\bar{\omega}$ are upper bounds of the variables, and $\bar{u}[t : t + T) = \{u(\tau) : t \le \tau < t + T\}$. The update rule of the continuous-time RL is given by

$$\delta(t) = r(t + T) + \frac{1}{T}\left((1 - \gamma T)\widehat{Q}(t + T) - \widehat{Q}(t)\right), \tag{8.45}$$

$$\dot{\theta}(t) = \alpha(t)\frac{\delta(t)}{T}\frac{\partial \widehat{Q}(t)}{\partial \theta(t)}, \tag{8.46}$$

where $\widehat{Q}(t + T) = \widehat{Q}(x(t + T), u(t + T); \theta(t))$ and $\widehat{Q}(t) = \widehat{Q}(x(t), u(t); \theta(t))$. The above update rule is also known as continuous-time (CT) critic learning (CL). The CT-CL method with RBFNNapproximation is as follows:

1. Solve the Q-function using the Q-function update law (8.46);
2. Update the control policy with $h^{j+1}(x(t)) = \arg\min_{u(t)}\widehat{Q}^{h^j}(x(t), u(t); \theta(t))$.

The standard actor-critic (AC) method can be written in continuous time, which is called continuous-time (CT) actor-critic learning (ACL). Here the V-function (8.19) is used instead of the Q-function (7.29). The policy is parameterized by the function approximator. The CT-ACL update rules are

$$\dot{\theta}(t) = \alpha_c(t)\frac{\delta(t)}{T}\frac{\partial \widehat{V}(t)}{\partial \theta(t)}, \tag{8.47}$$

$$\dot{\vartheta}(t) = \alpha_a(t)\Delta u(t)\frac{\delta(t)}{T}\frac{\partial \hat{h}(x(t); \vartheta(t))}{\partial \vartheta(t)}, \tag{8.48}$$

$$\delta(t) = r(t + T) + \frac{1}{T}\left((1 - \gamma T)\widehat{V}(t + T) - \widehat{V}(t)\right),$$

where $\alpha_c(t)$ and $\alpha_a(t)$ are the learning rates for the critic and actor, $\Delta u(t) \sim \mathcal{N}(0, \sigma^2)$ is random exploration noise, which plays as the persistent exciting (PE) signal. The value function approximation $V(t)$ satisfies the following relations.

$$\widehat{V}(t + T) = \widehat{V}(x(t + T); \theta(t)) = \Phi^\top(x(t + T))\theta$$
$$\widehat{V}(t) = \widehat{V}(x(t); \theta(t)) = \Phi^\top(x(t))\theta(t).$$

The critic is updated by (8.47), and the actor is updated by (8.48). The policy $h(x)$ is parameterized as

$$\hat{h}(x; \vartheta) = \Psi^\top(x)\vartheta,$$

where $\Phi(x)$ and $\Psi(x)$ are the vectors of basis functions of the critic and actor, respectively.

Note that these basis functions only depend on the state space. The CT-ACL has to approximate the critic with (8.47) and the actor with (8.48). We only approximate the critic as in (8.46), which is simpler than CT-ACL.

The main difference between discrete-time (DT) RL and the CT-CL is that DT-RL uses a one-step update but uses the interval $[t : t + T]$, which depends on the step size of the integrator. In each time step the Q function is updated taking into account previous Q functions in a small time interval.

The robustness of the CT-CL comes from the state-action value function. The CT-ACL method learns two separate functions: the value function and the control policy. The actor is updated using the TD error of the value function, while the critic is updated indirectly by the followed policy. Therefore, the critic depends on how well the actor learns the control policy. The value function of CT-CL has knowledge of the followed control policy, which helps to know how the control policy affects the system in a certain state.

8.5 Simulations and Experiments: Discrete-Time Case

We compare our modified reinforcement learning algorithms, large state (LS) - discrete action (DA) and LS - large action (LA), with various classical linear controllers and discontinuous controllers: LQR, PID, sliding mode control (SMC), and classical RL controllers, AC method (RL-AC)[13]. We discuss two cases: ideal control without uncertainty and robust control with worst-case uncertainty. To select the hyper-parameters of the RL methods, we use a random search method. We use prior knowledge to identify the range of the hyper-parameters. From the ranges, the parameters are selected at random until we find the best hyper-parameters.

Example 8.1 *Cart-Pole System*
Consider the cart-pole system given in Appendix A.4. The control objective is to balance the pole by moving the cart. In this simulation, the ideal parameters are $m = 0.1$ kg, $M = 1.0$ kg, and $l = 0.5$ m, The gravity is set to $g = 9.81$ m/s^2.

The worst-case uncertainties are: the parameters are increased by 100%, i.e., m is 0.2 kg, M is 2.0 kg, and l is 1.0 m, when $t > 5$ seconds. The cart position space is restricted as $x_c \in [-5, 5]$. The initial condition is $x_0 = [x_c, \dot{x}_c, q, \dot{q}]^\top = [0, 0, 0, 0.1]^\top$. We use RLs methods to learn the stabilizing controller. Each RL method has 1,000 episodes, and each episode has 2,000 steps. When our modified RL converges, the TD-error is modified to

$$\delta_t = r_{t+1} - Q(kNN_t, u_t). \tag{8.49}$$

Otherwise, the same TD error of Algorithms 8.1–8.2 is used. The learning hyper-parameters of the reinforcement learning methods are given in Table 8.1.

Since the state-action space is hued, the Q-function is approximated by

$$\widehat{Q}(x, u) = \widehat{\theta}^\mathsf{T} \Phi(x, u),$$

where $\widehat{\theta}$ is the weight vector, and $\Phi(x, u) : \mathbb{R}^n \to \mathbb{R}^N$ is the basis function. The update rule (8.13) for Q-learning is modified as

$$\widehat{\theta}_{t+1} = \widehat{\theta}_t + \alpha_t (r_{t+1} + \gamma \min_u (\Phi^\mathsf{T}(x_{t+1}, u)\widehat{\theta}_t) - \Phi^\mathsf{T}(x_t, u_t)\widehat{\theta}_t)\Phi(x_t, u_t).$$

We use normalized radial basis functions (NRBF) as the basis function. The parameters of NRBF are given in Table 8.1, where the first four elements stand for the cart-pole states, and the last one is the control input.

The classical AC method [31] approximates the state-value function and control policy as

$$\widehat{V}(x) = \widehat{\theta}_c^\mathsf{T} \Phi_1(x), \quad \widehat{h}(x) = \widehat{\theta}_a^\mathsf{T} \Phi_2(x), \tag{8.50}$$

where $\widehat{\theta}_c$ and $\widehat{\theta}_a$ are the weight vectors of the critic and the actor, respectively, and $\Phi_1(x)$ and $\Phi_2(x)$ are the basis functions, which only depend on states measures. The update rules of the classical AC method are given as follows:

$$\widehat{\theta}_{c,t+1} = \widehat{\theta}_{c,t} + \alpha_{c,t}(r_{t+1} + \gamma \Phi_1^\mathsf{T}(x_{t+1})\widehat{\theta}_{c,t} - \Phi_1^\mathsf{T}(x_t)\widehat{\theta}_c)\Phi_1(x_t),$$

$$\widehat{\theta}_{a,t+1} = \widehat{\theta}_{a,t} + \alpha_{a,t}(r_{t+1} + \gamma \Phi_1^\mathsf{T}(x_{t+1})\widehat{\theta}_{c,t} - \Phi_1^\mathsf{T}(x_t)\widehat{\theta}_c)\Delta u_{t-1}\Phi_2(x_t),$$

where α_c and α_a are the learning rate of the critic and the actor, respectively, and Δu_t is a zero-mean random exploration term. In this simulation $\Phi_1 = \Phi$.

Table 8.1 Learning parameters DT RL: cart pole system

Parameter	Q-learning	RL AC	RL LS-DA	RL LS-LA
α_t	0.3	-	0.09	0.3
$\alpha_{c,t}$	-	0.3	-	-
$\alpha_{a,t}$	-	0.05	-	-
γ	1.0	1.0	1.0	1.0
NRBF	$[10, 5, 10, 5, 5]^\mathsf{T}$	$[10, 5, 10, 5]^\mathsf{T}$	-	-
k	-	-	8	8

For the model-based robust control LQR [2], we use the following discrete algebraic Riccati equation to get the optimal solution:

$$A^T PA - P + S - A^T PB(B^T PB + R)^{-1} B^T PA = 0,$$

where $A \in \mathbb{R}^{4 \times 4}$ and $B \in \mathbb{R}^4$ are the linearized matrices of the dynamics. R and S are positive defined weight matrices. Here we know the dynamics (A.36), so we have the optimal control as

$$u_t^* = -Kx_t,$$

where $K = [3.1623, 28.9671, 3.5363, 3.7803]^T$.

The following SMC [8, 32] can stabilize the system if the sliding mode gain is bigger than the upper bound of all unknown dynamics,

$$u_t = K_m sign(Kx_t), \tag{8.51}$$

where K_m is the sliding gain. The sliding gain is chosen as $K_m = 10$. The PID control [33] has the form

$$u_t = K_p [x_c, q]^T + K_i \int_0^t [x_c, q]^T d\tau + K_d [\dot{x}_c, \dot{q}]^T.$$

They require a suitable tuning procedure [6]. We use the MATLAB control toolbox to tune the PID gains with the following values:

$$K_p = [5.12, 20.34]^T, K_i = [1.54, 0.57]^T, K_d = [1.51, 1.56]^T$$

Figure 8.1 shows the obtained results. After training, Figure 8.1(a) shows the control results of the ideal case. All controllers, LQR, PID, SMC, Q-learning, RL-AC, LS-DA, LS-LA, work well. Figure 8.1(b) shows the results of the disturbed case. Here we only increase the parameter values by 80% at time $t = 10$ s. We can see that LQR and PID are not stable.

SMC uses its chattering technique to cancel this disturbance and obtain a good result; however, the control gain is very large to compensate the disturbance. Our proposed RL has already learned the worst case (the parameters change 100%); it has good results and are also robust.

We also find that when the parameters increase by 90%, LA-DA and LS-LA can stabilize the system. But the LQR, PID, and SMC controllers are not stable. Here we can re-tune the PID and the SMC so that they can stabilize the system when the parameters increase by 90%. However, they become unstable in the ideal case (parameters without changes).

Both Q-learning and RL-AC are robust. Q-learning presents a large error compared with the other RL methods. RL-AC has better performance than Q-learning but has the overestimation problem in presence of the worst-case uncertainty, and it depends on the accuracy of the neural estimator. Figure 8.2 shows the mean errors, which are defined as $\bar{e} = \frac{1}{n} \sum_{i=1}^{n} e_i$. The blue bars represent the error of the

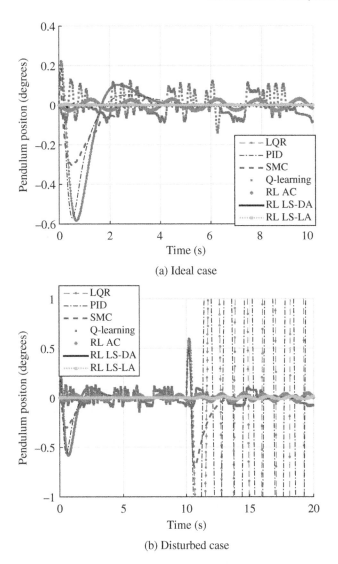

(a) Ideal case

(b) Disturbed case

Figure 8.1 Pole position.

RL methods at the ideal control case, and the red bars represent the error of the RL methods in presence of disturbances.

The traditional AC method has the overestimation problem for the action values for the worst-case learning. This problem affects the accuracy of the robust control. Figure 8.2 shows that the residual errors of the modified RL is much less than those of the traditional RL methods. Another reason is the AC method uses

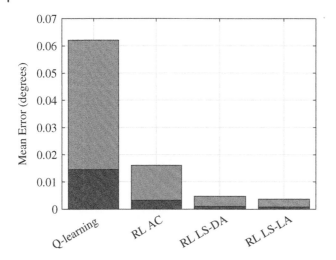

Figure 8.2 Mean error of RL methods.

two approximators for the value function and policy estimation. The approximation accuracies of these neural models strongly affect the performance of the RL method.

Finally, we show how γ affects the worst-case learning. The previous simulations use $\gamma = 1$ to learn the most recent uncertainties. Now we decrease γ to $\gamma = 0.8$, $\gamma = 0.5$, and $\gamma = 0.2$. In the learning phase, the parameters are also changed to 100%. However, robust controllers of our RL methods cannot stabilize the cart-pole system with an 80% change in parameters. When $\gamma = 0.8$, it can stabilize the cart system with a 55% change. $\gamma = 0.5$ can stabilize the cart system with a 42% change. $\gamma = 0.2$ can stabilize the system with a 26% change.

Example 8.2 *Experiments*

To further illustrate the effectiveness of the proposed methods, we use a 2-DOF planar robot (see Appendix A.3) as in Figure 8.3. The robot dynamics written as (8.1) is

$$\dot{x}_t = f(x_t) + g(x_t)u_t,$$

$$f(x_t) = \begin{bmatrix} 0 & 1 & 0 & 0 \\ 0 & 0 & 0 & 1 \\ 0 & 0 & 0 & 0 \\ 0 & 0 & 0 & 0 \end{bmatrix} x_t - \begin{bmatrix} 0 \\ 0 \\ M^{-1}(q)(C(q,\dot{q})\dot{q} + G(q)) \end{bmatrix},$$

$$g(x_t) = \begin{bmatrix} 0 \\ 0 \\ M^{-1}(q) \end{bmatrix}.$$

Figure 8.3 2-DOF planar robot.

Table 8.2 Learning parameters DT RL: 2-DOF robot

Parameter	Q-learning	RL LS-DA	RL LS-LA
α_t	0.5	0.4	0.1
γ	1.0	1.0	1.0
NRBF	$[20, 20, 10]^T$	-	-
k	-	9	8

The estimated parameters of the robot are $l_1 = 0.228$ m, $l_2 = 0.44$ m, $m_1 = 3.8$ kg, and $m_2 = 3$ kg, where l_i and m_i stand for the length and mass, respectively, of link i.

We only compare four methods: PID, Q-learning, LS-DA, and LS-LA. The desired positions are $q_d = \left[\frac{\pi}{3}, \frac{\pi}{4}\right]^T$ rad. We also add the following square wave in q_1 as the worst-case uncertainty, at $t = 10$ seconds,

$$\omega = 7sgn\left[\sin(3\pi t)\right] + 8\sin(5\pi t)$$

The learning hyper-parameters are given in Table 8.2, where the number of NRBF is for each DOF of the robot.

The number of episodes are 1,000. In each episode, there are 1,000 steps. With the same method as the cart-pole balancing, the final PID gains are: $K_p = diag\{90, 90\}, K_i = diag\{15, 15\}$, and $K_d = diag\{50, 50\}$. After the learning, we add the following perturbation at $t = 5$ seconds,

$$\omega = 5sgn\left[\sin(8\pi t)\right].$$

The comparison results are given in Figure 8.4. At the first 5 seconds all the controllers achieve the control task. After 5 seconds the perturbation is applied. Here the PID control cannot stabilize the joint angle q_1 and starts to oscillate. On the

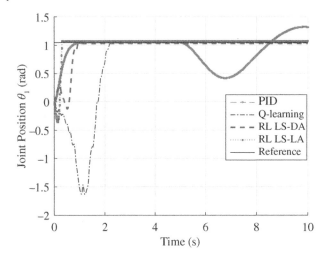

Figure 8.4 q_1 position regulation.

other hand, the three reinforcement learning methods are robust and achieve the control objective. For q_2 regulation, all controllers achieved the control task, since it doesn't have any perturbation.

Similarly to the cart-pole system, the gains of the PID control are tuned for the unperturbed case. The PID control is tuned to stabilize the perturbed system and guarantee the desired reference tracking. If the perturbed system is changed to the unperturbed system, then the PID control has poor performance because it has been tuned for the perturbed case. On the other hand RL methods overcome this issue since they are trained using the worst case uncertainty.

Discussion

The normal double-estimator technique [16] may not converge when the uncertainties are big. In this chapter we use the TD-error of Algorithms 8.1-8.2 to start a new control policy until it converges, and then it is changed as (8.49). This modified RL avoids the overestimation of the actions in certain states. Figure 8.5 shows the control actions of the cart-pole system at the upright position after 10 seconds. We can see the control policies of LS-DA and LS-LA do not overestimate the action values, compared with Q-learning and the RL-AC. Here RL-AC needs more time to learn the control policy.

Another technique used in this chapter is to use the previous control policy as a good experience. So the control signal is updated near the optimal policy, $u_t^* = \arg\min_u Q^*(kNN_t, u)$. Here the sub-optimal control uses the kNN, not the state x_t. The kNN approximation helps RL to learn a large state-action space with

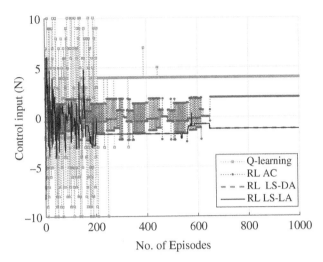

Figure 8.5 Control actions of the cart-pole balancing system after 10 seconds.

less computational effort. Compared with the other controllers such as LQR, PID, and SMC, the kNN-based RL is not only robust with respect the worst-case uncertainty but also gives good transient performances. The robust reward modification guarantees the convergence of the controller to the sub-optimal solution and also stability with respect to the disturbances.

The main problem of this approach is the design of the input state-action space, especially for the LS-DA case. Here the learning time and the computational effort increases since we have two estimators in parallel with a large input space. Here we assume that the worst case uncertainty or an upper bound is known in advance; otherwise it is required to estimate the uncertainty off-line. Also the correct choice of the hyper-parameters is a hard problem for any control task. To improve the controller, prior data information can be used as an initial condition for the value function to avoid start learning from scratch; also the use of a long short-term memory such as eligibility traces [34] can accelerate the convergence. Optimization procedures can be used for the kNN rule to find the optimal number of k-nearest neighbors.

8.6 Simulations and Experiments: Continuous-Time Case

To evaluate the performance of the RL approach, we apply two benchmark problems. We compare our continuous-time critic learning (CT-CL) with the $\mathcal{H}_2$ solution (LQR), and the continuous-time actor-critic learning (CT-ACL) method [35].

Table 8.3 Learning parameters CT RL: cart pole system

Parameters	CT-CL	CT-ACL
No. of RBFs	$[10, 5, 10, 5, 5]^\mathsf{T}$	$[10, 5, 10, 5]^\mathsf{T}$
Learning rate $\alpha(t)$	0.3	-
Discount factor γ	1	1
Critic learning rate $\alpha_c(t)$	-	0.3
Actor learning rate $\alpha_a(t)$	-	0.05

Example 8.3 *Simulations*

Consider the cart-pole system (see Appendix A.4). CT-RL and CT-ACL are designed to satisfy (8.44), where they learn to stabilize the pendulum at the upright position using the worst-case uncertainty. The worst-case uncertainty is $\omega = 5\sin(\pi t)$. The learning parameters and number of RBFs are given in Table 8.3.

For the CT-CL, we have five dimensions instead of four because the control input space or action space is added. For simplicity the basis functions of the actor and critic are the same, i.e., $\Phi(x) = \Psi(x)$. The initial condition is $x_0 = (x_c, \dot{x}_c, q, \dot{q})^\mathsf{T} = (0, 0, 0.2, 0.1)^\mathsf{T}$. This is the most difficult position to control.

The RL algorithm learns successfully from this starting point, and it can stabilize the other initial points. We use 1,000 episodes to train the RL algorithm. In each episode, there are 1,000 steps. For the exploration phase a random Gaussian noise is added to the control input in the first 300 episodes.

After the controller is trained by the reinforcement learning methods, we compare them with the LQR solution of the $\mathcal{H}_2$ problem. First we linearize the system at the point $x = (0, 0, 0, 0)^\mathsf{T}$, which yields

$$\dot{x}(t) = Ax(t) + B(u(t) + \omega(t)), \tag{8.52}$$

where $A \in \mathbb{R}^{4\times4}$ and $B \in \mathbb{R}^4$ are the linearized matrices of (A.36). From (8.52) we can see that the disturbance is coupled with the control input. We use the following algebraic Riccati equation (ARE) to obtain the matrix $P \in \mathbb{R}^{4\times4}$ for the optimal solution:

$$A^\mathsf{T}P + PA - P + S - PBR^{-1}B^\mathsf{T}P = 0,$$

where $S = I$ and $R = 0.1$ are the weight matrices of the $\mathcal{H}_2$ control index (8.33). Since the optimal cost function can be expressed as

$$V^*(x(t)) = x^\mathsf{T}(t)Px(t),$$

for some kernel matrix $P \in \mathbb{R}^{4\times4}$, the optimal control (8.38) is

$$u(t)^* = -R^{-1}B^\mathsf{T}Px(t).$$

Figure 8.6 Pole position.

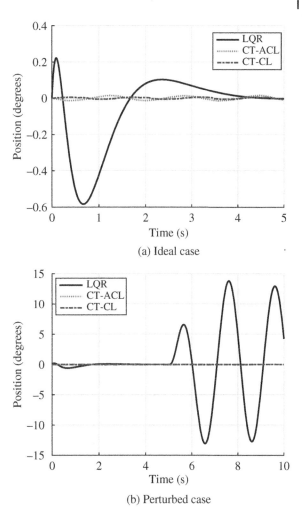

(a) Ideal case

(b) Perturbed case

The simulation consists of two parts: when $t \leq 5$ there is no disturbance, i.e., $\omega = 0$, which is the ideal case; and when $t > 5$ seconds, we apply a disturbance of $\omega = 4\sin(\pi t)$, which is the perturbed case.

The solution of the ideal case is shown in Figure 8.6(a). Both LQR and RL algorithms work well and stabilize the pendulum position in the first 5 seconds. The solution of the perturbed case is shown in Figure 8.6(b). The LQR solution in presence of the disturbance is unstable and cannot stabilize the pendulum position, On the other hand, the RL algorithms are stable and robust because they have learned the worst case, i.e., $\omega = 5\sin(\pi t)$.

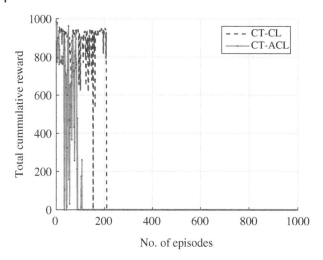

Figure 8.7 Total cumulative reward curve.

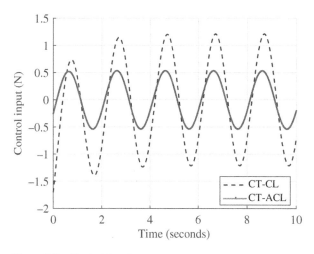

Figure 8.8 Control input.

Figure 8.7 shows the total cumulative reward. We can observe that the CT-ACL minimizes the reward in fewer episodes compared to the CT-CL. From the reward plot we obtain the number of failures (number of times the pendulum falls) of each RL method. For the CT-ACL these are 87 failures, which represent 8.7% of the total number of episodes. On the other hand, the CT-CL has 208 failures, which represent 20.8% of the total number of episodes. This result is not favorable for the approach.

Figure 8.9 *Q*-function learning curves for $\dot{x}_c = \dot{q} = 0$.

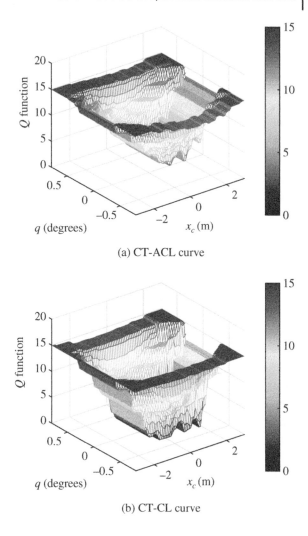

(a) CT-ACL curve

(b) CT-CL curve

Figure 8.8 shows the control input of each RL method. The results show that the CT-ACL has a smaller control input in comparison to the CT-CL. As we mentioned in the discrete-time case, the use of the min operator to obtain the best control policy h may cause the overestimation problem, which means that the controller uses high action values for states that do not require them.

However, the overestimation problem is beneficial for the continuous-time robust control problem. Figure 8.9 gives one of the Q-function surfaces between the cart position x_c and the pendulum position q when the cart and pendulum velocity are zero, i.e., $\dot{x}_c = \dot{q} = 0$. The CT-ACL curve shows the robust Q function;

remember that for a fixed control policy $h(x(t))$ the value function satisfies

$$V(x(t)) = Q(x(t), h(x(t))),$$

where the pendulum is stabilized in a certain cart position interval. The CT-CL curve shows a different robust Q function where the interval of the cart position is larger than for the CT-ACL method. This larger interval means that the cart can stabilize the pendulum in different cart positions and keep the robustness property. We use the following integral squared error (ISE) to show the robustness of the continuous-time RL

$$ISE = \int_{t_0}^{t} (\kappa \tilde{q})^2 d\tau,$$

where $\tilde{q} = -q$, and κ is a scaling factor. The integrator of the ISE is reset at each change of sign of the disturbance, i.e., $reset = sign(\omega)$. It uses a scaling factor of $\kappa = 100$, and the results are given in Figure 8.10.

From the ISE plot we can see that for the CT-ACL method, the pendulum position error increases when the cart moves away from the near-optimal, robust value; however, the pendulum is still stabilized. On the other hand, the CT-CL shows that even if the cart moves the position error does not increase because the obtained control policy is more robust.

Example 8.4 *Experiments*
To further illustrate the effectiveness of the CT approach, we use a 2-DOF planar robot (see Figure 8.3 and Appendix A.3) and compare the CT-CL approach with the LQR solution and the CT-ACL method. The desired position is $q_d = \left[\frac{3\pi}{4}, \frac{\pi}{4}\right]^T$ rad. The control objective is to force the two joint angles q_1 and q_2 to the desired position q_d.

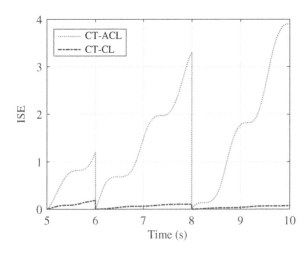

Figure 8.10 ISE comparisons.

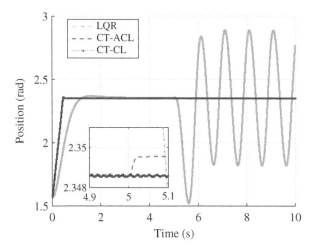

Figure 8.11 Joint position q_1 tracking.

Table 8.4 Learning parameters CT RL: 2-DOF robot

Parameters	CT-CL	CT-ACL
No. of RBFs	$[20, 10, 10]^\top$	$[20, 10]^\top$
Learning rate $\alpha(t)$	0.3	-
Discount factor γ	1	1
Critic learning rate $\alpha_c(t)$	-	0.3
Actor learning rate $\alpha_a(t)$	-	0.1

The parameter tuning was done by iteration until the best performance was obtained. The optimal learning parameters and number of RBFs are given in Table 8.4.

Since we have to control two joint angles, we need two Q functions, denoted by $\hat{Q}_1$ and $\hat{Q}_2$. Each approximate Q-function are used for the design of the robust control policy. For each Q-function we used the same number of RBFs and learning parameters for both the CT-CL and the CT-ACL. The number of episodes are 100, with 1,000 steps per episode. A bounded white noise term between −1 and 1 is added to the control inputs as an exploration term for the first 50 episodes. The RL methods learn the control objective using the following worst disturbance applied at the first robot link: $\omega = 30\sin(2\pi t)$.

After the controller is trained by the reinforcement learning methods, we compare them with the LQR solution of the $\mathcal{H}_2$ problem. The robot dynamics are linearized at the point $x = (0, 0, 0, 0)^\top$. The linearized dynamics have the form given

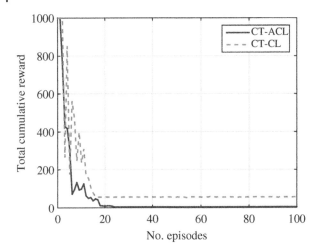

Figure 8.12 Total cumulative reward.

by (8.52), where $A \in \mathbb{R}^{4 \times 4}$, $B \in \mathbb{R}^{4 \times 2}$ are the linearized matrices of the planar robot. From the linearized model the $\mathcal{H}_2$ controller is obtained by solving the ARE (6.11) using the weight matrices $S = I$ and $R = 0.1I$.

A similar procedure to the cart-pole balancing problem is carried out in this experiment. When $t \leq 5$ seconds, there is no disturbance, i.e., $\omega = 0$, which is the ideal case; and when $t > 5$ seconds, we apply a disturbance of $\omega = 20\sin(2\pi t)$, which is the perturbed case. The comparison of results is given in Figure 8.11. The LQR solution is unstable when the disturbance is applied, while the CT-ACL and CT-CL methods are still stable. Here both methods have good robustness with small position error differences. For q_2 all the controllers are stable and achieve the control objective.

Figure 8.12 shows the total cumulative reward. Here we can see that the CT-ACL minimizes the reward in a better way than the CT-CL. Since the CT-CL has an overestimation problem, the control input is bigger than the control input of the CT-ACL, and hence the reward is increased. Nevertheless, as it is shown in Figure 8.11, the robust performance is achieved and the position error is minimized.

The learning curves of the approximate value functions $\hat{Q}_1$ and $\hat{Q}_2$, for either the CT-CL or CT-ACL, are shown in Figure 8.13. We can observe that the minimum of the value functions is located at the origin, which demonstrate the position tracking results.

From the previous benchmark problems we can see that the robust reward modification can guarantee the convergence of the RL methods to a near-optimal

Figure 8.13 Q-function learning curves for $\dot{x}_c = \dot{q} = 0$.

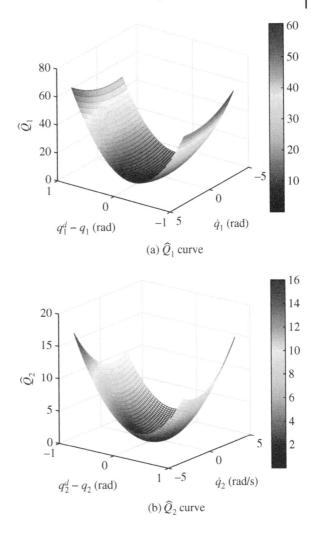

(a) $\hat{Q}_1$ curve

(b) $\hat{Q}_2$ curve

solution (when there is no disturbance) and robust responses in the presence of disturbances.

The results of the simulation and experiment show that the Q-function and control policy of the continuous-time RL methods are robust without knowledge of the system dynamics. Compared to the actor-critic algorithm, our approach uses the actions to update the value function. This method gives the knowledge of actions to certain states. The obtained results show that our approach overestimates the actions and therefore increases the total cumulative reward. It is favorable because it assures good control policy in the presence of disturbances.

8.7 Conclusions

This chapter presents a simple and robust control method using reinforcement learning in both discrete and continuous time. It uses two different approximators based on the k-nearest neighbors and NRBF. The first one uses the double estimator technique to avoid overestimation of the action values, while the second one uses an actor-critic algorithm, which has a similar function. This H_2/H_∞ control is robust under the worst-case uncertainty. The convergence of the RL methods using the kNN rule is analyzed, while the convergence of the continuous time RL is the same as in the previous chapter. The algorithms are compared with the classical linear and nonlinear controllers. The results of the simulation and the experiment show that the RL algorithms are robust with sub-optimal control policy with respect to the worst-case uncertainty.

References

1 C. Vivas and F. Rubio, "Nonlinear H_∞ measurement feedback control of Euler Lagrange Systems," *IFAC Proceedings Volumes (IFAC-PapersOnline)*, pp. 391–396, 2005.

2 L. Márton, A. Scottedward Hodel, B. Lantos, and Y. Hung, "Underactuated robot control: Comparing LQR, subspace stabilization, and combined error metric approaches," *IEEE Transactions on Industrial Electronics*, vol. 55, no. 10, pp. 3724–3730, 2008.

3 P. Millán, L. Orihuela, G. Bejarano, C. Vivas, T. Alamo, and F. Rubio, "Design and application of suboptimal mixed H_2/H_∞ controllers for networked control systems," *IEEE Transactions on Control Systems Technology*, vol. 20, no. 4, pp. 1057–1065, 2012.

4 M. Sayahkarajy and Z. Mohamed, "Mixed sensitivity H_2/H_∞ control of a flexible-link robotic arm," *International Journal of Mechanical and Mechatronics Engineering*, vol. 14, no. 1, pp. 21–27, 2014.

5 P. A. Ioannou and J. Sun, *Robust Adaptive Control*. Dover: Mineola, New York, 2012.

6 J. G. Romero, A. Donaire, R. Ortega, and P. Borja, "Global stabilisation of underactuated mechanical systems via PID passivity-based control.," *IFAC-PapersOnLine*, 2017.

7 H. Zhang, Y. Shi, and A. Mehr, "Robust H_∞ PID control for multivariabl networked control systems with disturbance/noise attenuation," *International Journal of Robust and Nonlinear control*, vol. 22, no. 2, pp. 183–204, 2012.

8 R. Xu and U. Özgüner, "Sliding mode control of a class of underactuated systems.," *Automatica*, vol. 44, 2014.

9 Y. Zuo, Y. Wang, X. Liu, S. Yang, L. Huang, X. Wu, and Z. Wang, "Neural network robust H_∞ tracking control strategy for robot manipulators," *Applied Mathematical Modelling*, vol. 34, no. 7, pp. 1823–1838, 2010.

10 Q. Chen, L. Shi, J. Na, X. Ren, and Y. Nan, "Adaptive echo state network control for a class of pure-feedback systems with input and output constraints," *Neurocomputing*, vol. 275, pp. 1370–1382, 2018.

11 H. Zhang, L. Cui, X. Zhang, and Y. Luo, "Data-driven robust approximate optimal tracking control for unknown general nonlinear systems using adaptive dynamic programming method," *IEEE Transactions on Neural Networks*, vol. 33, no. 12, pp. 2226–2236, 2011.

12 F. Lewis, D. Vrable, and K. Vamvoudakis, "Reinforcement learning and feedback control using natural decision methods to desgin optimal adaptive controllers," *IEEE Control Systems Magazine*, 2012.

13 I. Grondman, M. Vaandrager, L. Buşoniu, R. Babûska, and E. Schuitema, "Efficient model learning methods for actor-critic control," *IEEE Transactions on Systems, man, and cybernetics. Part B: Cybernetics*, vol. 42, no. 3, 2012a.

14 M. Palanisamy, H. Modares, F. L. Lewis, and M. Aurangzeb, "Continuous-time Q-learning for infinite-horizon discounted cost linear quadratic regulator problems," *IEEE Transactions on Cybernetics*, 2014.

15 M. Ganger, E. Duryea, and W. Hu, "Double sarsa and double expected sarsa with shallow and deep learning," *Journal of Data Analysis and Information Processing*, 2016.

16 H. van Hasselt, "Double Q-learning," *In Advances in Neural Information Processing Systems (NIPS)*, pp. 2613–2621, 2010.

17 L. Buşoniu, R. Babûska, B. De Schutter, and D. Ernst, *Reinforcement learning and dynamic programming using function approximators.* CRC Press, Automation and Control Engineering Series, 2010.

18 J. A. Martín H., J. de Lope, and D. Maravall, "*The kNN-TD reinforcement learning algorithm,*" *Springer-Verlag*, J. Mira *et al.* (eds): IWINAC 2009, Part I, LNCS 5601, pp. 305–314, 2009.

19 J. A. Martín H. and J. de Lope, "Ex$\langle a \rangle$: An effective algorithm for continuous actions reinforcement learning problems," *Industrial Electronics*, 2009.

20 J. Škach, B. Kiumarsi, F. Lewis, and O. Straka, "Actor-critic off-policy learning for optimal control of multiple-model discrete-time systems," *IEEE Transactions on Cybernetics*, vol. 48, no. 1, pp. 29–40, 2018.

21 B. Luo, H.-N. Wu, T. Huang, and D. Liu, "Data-based approximate policy iteration for affine nonlinear continuous-time optimal control design," *Automatica*, vol. 50, no. 12, pp. 3281–3290, 2014.

22 I. Grondman, M. Vaandrager, L. Buşoniu, R. Babûska, and E. Schuitema, "Actor-critic control with reference model learning," *Proc. of the 18th World*

Congress The International Federation of Automatic Control, pp. 14723–14728, 2011.

23 K. Vamvoudakis and F. Lewis, "Online actor-critic algorithm to solve the continuous-time infinite horizon optimal control problem," *Automatica*, vol. 46, no. 5, pp. 878–888, 2010.

24 B. Kiumarsi, K. G. Vamvoudakis, H. Modares, and F. L. Lewis, "Optimal and autonomous control using reinforcement learning: A survey," *IEEE Transactions on Neural Networks and Learning Systems*, vol. 29, no. 6, 2018.

25 A. Al-Tamimi, F. Lewis, and M. Abu-Khalaf, "Model-free Q-learning designs for linear discrete-time zero-sum games with application to H_∞ control," *Automatica*, vol. 43, no. 3, pp. 473–481, 2007.

26 B. Kiumarsi, F. Lewis, and Z. Jiang, "H_∞ control of linear discrete-time systems: Off policy reinforcement learning," *Automatica*, vol. 78, pp. 144–152, 2017.

27 L. Xiao, K. Li, and M. Duan, "Computing time-varying quadratic optimization with finite time convergence and noise tolerance: A unified framework for zeroing neural network," *IEEE Transactions on Neural Networks and Learning Systems*, pp. 1–10, 2019.

28 T. Cover and P. Hart, "Nearest neighbor pattern classification," *IEEE Transactions on Information Theory*, vol. IT-13, pp. 21–27, 1967.

29 S. Sun and R. Huang, "An adaptive *k*-nearest neighbor algorithm," *2010 Seventh International Conference on Fuzzy Systems and Knowledge Discovery*, 2010.

30 A. Perrusquía and W. Yu, "Robust control under worst-case uncertainty for unknown nonlinear systems using modified reinforcement learning," *International Journal of Robust and Nonlinear Control*, 2020.

31 I. Grondman, L. Buşoniu, and R. Babûska, "Model learning actor-critic algorithms: performance evaluation in a motion control task," *51st IEEE Conference on Decision and Control (CDC)*, pp. 5272–5277, 2012b.

32 W. Yu and A. Perrusquía, "Simplified stable admittance control using end-effector orientations," *International Journal of Social Robotics*, 2019.

33 A. Perrusquía and W. Yu, "Human-in-the-loop control using euler angles," *Journal of Intelligent & Robotic Systems*, 2019.

34 A. Perrusquía, W. Yu, and A. Soria, "Position/force control of robots manipulators using reinforcement learning," *Industrial Robot: the international journal of robotics research and application*, vol. 46, no. 2, pp. 267–280, 2019.

35 K. Doya, "Reinforcement learning in continuous time and space," *Neural Computation*, vol. 12, no. 1, pp. 219–245, 2000.

9

Redundant Robots Control Using Multi-Agent Reinforcement Learning

9.1 Introduction

Task-space control is an important job in robotics and robot control, where the robot end effector is forced to reach the desired reference position and orientation in the workspace [1]. Since the control action is in joint space, classic control approaches require the inverse kinematics or velocity kinematics (Jacobian matrix). Redundant robots have more joint angles than the degrees of freedom (DOFs) of the end effector [2]. These extra DOFs help the robot to achieve more complex tasks, such as the tasks performed by the human arm. It is very difficult to obtain an accurate representation of the inverse kinematics and the Jacobian matrix for redundant robots [3] because there exist multiple solutions [4, 5]. The classical kinematic methods are developed for non-redundant robots [6, 7].

There exist several solutions to the kinematic problem, where the most famous approaches are in velocity level [8] using the Jacobian matrix. For a redundant robot the Jacobian is not a square matrix, so the space transformation from task to joint space cannot be applied directly. The existing methods are based on the pseudo-inverse method, which could have controllability problems if the robot is at a singular point [9–11]. Approximate solutions, as the singularity avoidance method, are used to solve the controllability problem. They have high or very small joint rates, which affects the behavior of the controller [2]. Some approaches use augmented models that use components of the end-effector orientation [12]. It requires the dynamic model of the robot.

Learning algorithms [13], such as neural networks, are used to solve the inverse kinematic problem in [14]. These methods learn a specific trajectory [15, 16] or the kinematic model [17]. Reinforcement learning (RL) has been used for robot

Human-Robot Interaction Control Using Reinforcement Learning, First Edition. Wen Yu and Adolfo Perrusquía.
© 2022 The Institute of Electrical and Electronics Engineers, Inc. Published 2022 by John Wiley & Sons, Inc.

control [18, 19] based on model-free and online algorithms [20]. RL methods for robot control include three subclasses [18, 21, 22]:

1) Value iteration: The controller obtains the optimal policy by taking the minimal return from every value or action value function, e.g., Q-learning algorithm [21].
2) Policy iteration: The controller evaluates its policies by constructing their value functions and use them to improve their policies, e.g., Sarsa [20], LSTD-Q algorithm.
3) Policy search: The controllers use optimization techniques to directly search for an optimal policy, e.g., policy gradient and actor-critic methods [18, 22].

The value and policy iteration algorithms require expensive optimization procedures to estimate the value function, while policy search algorithms store the value and the policy. The control actions can be calculated directly from the learned policy. Therefore, policy gradient and actor-critic methods are preferred for robot control [23].

The above RL methods are designed for 1-DOF. If the robot has a high number of DOFs, the space dimension and the computational load of the RL algorithm are increased [24, 25]. Multi-agent reinforcement learning (MARL) can deal with high DOFs [26], but these approaches require another kind of methodology to deal with the high state-action space such as function approximators. The main problem of the MARL for robot control is the curse of dimensionality, which is caused by the exponential growth of the discrete state-action space [27]. To solve the dimensionality problem, several architectures are designed, such as policy gradient [28] and actor-critic methods [29], where the value function and the policy are represented by differentiable parameterization [30]. However, these representations cannot be applied directly for redundant robots because the space dimension is still large [31]. So the control problems of redundant robots in task-space are

1) The accurate representation of the inverse kinematics solution is not available, because the inverse kinematics solution for redundant robots has multiple solutions.
2) The solution of the Jacobian matrix is not available, because the Jacobian matrix is not a square matrix and loses rank at singular points.

We will first discuss the classical joint and task space control schemes, which give the theory and the first solution of the inverse and velocity kinematics problems. Then we give the reinforcement learning approach, which is inspired in the classical solutions.

9.2 Redundant Robot Control

To obtain the task-space control, we first need the relation between the joint angles $q \in \mathbb{R}^n$ and the end effector $x \in \mathbb{R}^m$. They are given in Appendix A (see (A.1) and (A.2)):

$$x = f(q), \quad q = invk(x), \tag{9.1}$$

where $f(\cdot) : \mathbb{R}^n \to \mathbb{R}^m$ is the forward kinematics, and $invk(\cdot) : \mathbb{R}^m \to \mathbb{R}^n$ the inverse kinematics. For the redundant robot $n > m$, the inverse kinematics $invk(\cdot)$ is not feasible. To avoid this problem, the velocity kinematics is applied. It is expressed as (A.3)

$$\dot{x} = \frac{\partial f(q)}{\partial q} \dot{q} = J(q)\dot{q}, \tag{9.2}$$

where $J(q) \in \mathbb{R}^{n \times m}$ is the Jacobian matrix.

The Jacobian matrix has two associated subspaces:

1) The range space: the range of the Jacobian matrix is given by the transformation mapping

$$\mathfrak{R}[J(q)] = \{J(q)\dot{q} | \dot{q} \in \mathbb{R}^n\}.$$

2) The null space: the null space of the Jacobian is given by the following transformation mapping

$$\mathfrak{N}[J(q)] = \{\dot{q} \in \mathbb{R}^n | J(q)\dot{q} = 0\}.$$

Redundant robots have null space because some joint velocities do not generate any velocity at the end effector, but they generate joint motions.

There are two methods for task-space control: task-space design and joint-space design (see Fig. 9.1). The task-space design needs the inverse of the Jacobian J^+. If we use PID control,

$$\tau = J^+ \left(K_p e - K_d \dot{x} + K_i \int_0^t e(s)ds \right), \tag{9.3}$$

where $K_p, K_d, K_i \in \mathbb{R}^{m \times m}$ are the proportional, derivative, and integral positive definite diagonal matrices gains, respectively, and $e = x_d - x$ is the Cartesian position error.

The joint-space design needs the inverse kinematics $invk(x) = f^{-1}$. The joint-space PID control law is

$$\tau = K_p e_q - K_d \dot{q} + K_i \int_0^t e_q(s)ds, \tag{9.4}$$

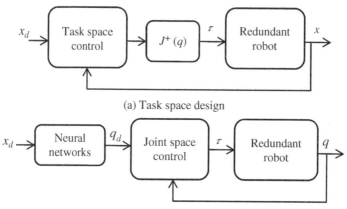

(a) Task space design

(b) Joint-space design with neural networks

Figure 9.1 Control methods of redundant robots.

where $e_q = q_d(x_d) - q$ is the joint position error, and $q_d(x_d)$ is obtained by the inverse kinematics, which is approximated by the neural networks. Here $K_p, K_d, K_i \in \mathbb{R}^{n \times n}$.

Task-Space Design

For redundant robots, the Jacobian matrix is not square. The simplest method to obtain the inverse of the Jacobian is to use the Moore-Penrose pseudoinverse:

$$J^\dagger = J^\top \left(J J^\top \right)^{-1}. \tag{9.5}$$

If the Jacobian J is not full rank, the singular-value-decomposition (SVD) can approximate the pseudo-inverse:

$$J^\dagger = v \sigma^* w^\top, \tag{9.6}$$

where σ^*, v, and w are the singular decomposition matrices.

The main problem of these pseudoinverse methods is that they do not exploit the velocities in the null space, so they require additional tasks, which are specified as [2]

$$\dot{q} = \dot{q}_\mathfrak{R} + \dot{q}_\mathfrak{N},$$

where the velocity in the null space is obtained by

$$\dot{q}_\mathfrak{N} = (I - J^\dagger J)v,$$

for the arbitrary $v \in \mathbb{R}^n$ [8].

In the augmented Jacobian method [10], an additional task is defined as:

$$x_a = l(q),$$

where $x_a \in \mathbb{R}^{n-m}$ is the vector defined by the function $l(\cdot)$, which depends on the joint angles. The augmented task space is expressed as $z = [x, x_a]^T$, so the new Jacobian satisfies

$$\dot{z} = \begin{bmatrix} J \\ L \end{bmatrix} \dot{q} = J_a \dot{q}, \quad L = \frac{\partial l(q)}{\partial q} \in \mathbb{R}^{(n-m) \times n}, \tag{9.7}$$

where $J_a \in \mathbb{R}^{n \times n}$ is a square matrix, and $\dot{q} = J_a^{-1} \dot{z}$. However, the augmented task space may bring extra singularities, and J_a^{-1} may not exist at all times.

We use the following optimization methods. The cost function is

$$J = \|J\dot{q} - \dot{x}_d\|^2, \tag{9.8}$$

where x_d is the desired position. To avoid singularities, the weighted norm of the joint rates is added to the cost function,

$$J = \|J\dot{q} - \dot{x}_d\|^2 + \|\xi\dot{q}\|^2, \tag{9.9}$$

where ξ is a penalization constant. The solution of the optimization problem (9.9) is

$$\dot{q} = \left(J^T J + \xi^2 I \right)^{-1} J^T \dot{x}_d. \tag{9.10}$$

The solution (9.10) is unique and close to the exact values. It avoids infinite high joint rates due to the singularities.

Remark 9.1 *The pseudoinverse method does not exploit the velocities in the null space. The SVD method is an off-line method that loses rank in presence of singularities. The augmented Jacobian incorporates new kinematic dynamics, which produces a new singularity. The optimal method can deal with some kinematic problems by choosing ξ. However, a small ξ at the singularity point gives big rate velocities, but a big ξ yields small rate velocities. On the other hand, any task-space control discussed in Chapter 4 can be used for this approach.*

Joint-Space Design

Joint-space control, shown in Figure 9.1, uses neural networks (NN) to learn the inverse kinematics instead of the Jacobian matrix. The NN scheme is a single hidden layer feedforward neural network, shown in Figure 9.2. It has n outputs (joint angles) and 6 inputs, which represent the robot end-effector pose (X, Y, Z) and orientation (α, β, γ).

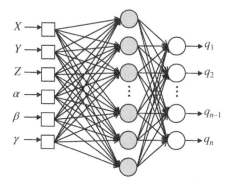

Figure 9.2 One hidden layer feedforward network.

Using both the position and orientation of the end effector helps to avoid multiple solutions of the inverse kinematics [5]. Because the joint space can be considered as the inverse mapping of the Cartesian space and vice versa, the position and orientation components x are the input and the joint angles q are the output of the neural network.

Each joint output of the NN with L hidden nodes is

$$q_i(x) = \sum_{j=1}^{L} W_j G(a_j, b_j, x) \quad i = 1, 2, ..., n, \tag{9.11}$$

where a_j and b_j are the learning parameters of the hidden nodes, W_j is the weight that connects the jth node to the output node, and $G(a_j, b_j, x)$ is the output of the jth hidden nodes with respect to the input x.

We use an additive hidden node where the activation function $g(x)$ is a hyperbolic tangent sigmoid function, and $G(a_j, b_j, x)$ is defined by

$$G(a_j, b_j, x) = g(a_j \cdot x + b_j). \tag{9.12}$$

For a given set of training samples $(x_j, t_j)_{j=1}^{N} \subset \mathbb{R}^n \times \mathbb{R}^m$, the output is expected to be equal to the targets:

$$q_i(x) = \sum_{j=1}^{L} W_j G(a_j, b_j, x_s) = t_s, \quad s = 1, 2, ..., N. \tag{9.13}$$

The above expression can be expressed as

$$HW = T, \tag{9.14}$$

where

$$H = \begin{bmatrix} G(a_1, b_1, x_1), & \cdots, & G(a_L, b_L, x_1) \\ \vdots, & \cdots, & \vdots \\ G(a_1, b_1, x_N), & \cdots, & G(a_L, b_L, x_N) \end{bmatrix}_{N \times L}, \tag{9.15}$$

$$
W = \begin{bmatrix} W_1^\top \\ \vdots \\ W_L^\top \end{bmatrix}_{L \times m}, \quad T = \begin{bmatrix} t_1^\top \\ \vdots \\ t_N^\top \end{bmatrix}_{N \times m}.
\tag{9.16}
$$

The weight W is obtained by the least square solution, i.e., min $\|W\|$ that minimizes $\|HW - T\|$, which yields the following solution using the Moore-Penrose pseudoinverse

$$
\hat{W} = H^\dagger T.
\tag{9.17}
$$

Remark 9.2 *The joint-space controller as in Section 4.3 requires that NN gives a good approximation of the inverse kinematics to achieve the desired task-space position. A wrong inverse kinematics may lead to wrong joint angles and wrong poses. To obtain reliable solutions, the NN needs a big training data and a long learning time.*

9.3 Multi-Agent Reinforcement Learning for Redundant Robot Control

Reinforcement learning (RL) for robot control can be regarded as a Markov decision process (MDP) [21]. The MDP is a tuple $\langle \mathcal{X}, \Delta Q, f, \rho \rangle$, where $\mathcal{X}$ is the state space of the task space components, ΔQ is the action space, which is represented by small joint displacements. The state transition is determined by the forward kinematics,

$$
f : \mathcal{X} \times \Delta Q \to \mathcal{X}.
\tag{9.18}
$$

The scalar reward function is

$$
\rho : \mathcal{X} \times \Delta Q \to \mathbb{R}.
$$

The forward kinematics (9.18) returns the new state $x_{t+1} \in \mathcal{X}$ of the task-space components after applying a small joint displacement $\Delta q \in \Delta Q$ at the current joint state q. After the transition, the RL algorithm receives the scalar reward as

$$
r_{t+1} = \rho(x_t, q_t + \Delta q_t),
$$

which evaluates the immediate effect of the displacement Δq, namely the transition from x_t to x_{t+1}. The joint displacements are chosen according to the policy

$$
q(x) : \mathcal{X} \to \Delta Q.
$$

The proposed learning scheme using RL is shown in Figure 9.3. Here the control actions are the joint displacement $\Delta q = [\Delta q_1, \cdots, \Delta q_r]^\top$, where $\Delta q_i, (i = 1, \cdots, r)$ are discrete displacements. The RL algorithm uses the joint

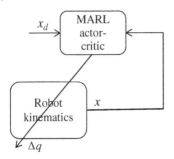

Figure 9.3 RL control scheme.

space displacements to update the robot kinematics until the robot achieves the desired position x_d. The update law of the forward kinematics (9.1) is

$$x = f(q + \Delta q) = f(\sigma), \qquad (9.19)$$

where $\sigma = q + \Delta q$. The objective of our RL is to find the inverse kinematics solution and force the end effector to reach certain desired position in task space, such that the discounted sum of future rewards is minimized.

Let's define the value function as

$$V^{q(x)}(x) = \sum_{i=t}^{\infty} \gamma^{i-t} \rho(x, \sigma) \text{ with } x_t = x, \qquad (9.20)$$

where $\gamma \in (0, 1]$ is the discount factor. (9.20) satisfies the following Bellman equation:

$$V^{q(x)}(x) = \rho(x, q(x)) + \gamma V^{q(x)}(x_{t+1}), \qquad (9.21)$$

where $x_{t+1} = f(x_t, q(x_t))$ is the forward kinematics of (9.19). The optimal value function satisfies

$$V^*(x) = \min_{q(x)} V^{q(x)}(x).$$

The inverse kinematics is given by

$$q^*(x) = \arg\min_{\sigma} \left[\rho(x, \sigma) + \gamma V^*(x_{t+1}) \right]. \qquad (9.22)$$

The value function can be written as the action value function by the following Bellman equation:

$$Q^{q(x)}(x, \sigma) = \rho(x, \sigma) + \gamma Q^{q(x)}(x_{t+1}, q(x_{t+1})). \qquad (9.23)$$

The optimal action-value function satisfies

$$Q^*(x, \sigma) = \min_{q(x)} Q^{q(x)}(x, \sigma).$$

Finally, the inverse kinematics solution is

$$q^*(x) = \arg\min_{\sigma} Q^*(x, \sigma). \qquad (9.24)$$

In (9.22) the Q-function depends on the joint displacements, while the V-function describes the task-space states. It is easier to use V-functions than Q-functions, since the Q-functions depend both on x and Δq and need the forward kinematics transition [32]. The Bellman equations (9.21) and (9.23) are the baseline equations to design RL algorithms. They are the tabular methods for a single DOF [21].

Since redundant robots have many joint DOFs, the single DOF RL cannot be applied, because there are too many tables to store all possible combinations of the joint angles. Also the components of the desired position are coupled. It is impossible to use separate reward functions for each DOF in the task space.

We consider each DOF as an agent and use multi-agent reinforcement learning (MARL) [18, 33]. The MDP is the following stochastic game:

$$MDP = \langle \mathcal{X}, \Delta Q_1, ..., \Delta Q_n, f, \rho_1, ..., \rho_n \rangle,$$

where n is the number of the joints, $\Delta Q_i, i = 1, ..., n$, are the action space (joint displacements), and $\rho_i : \mathcal{X} \times \Delta Q \to \mathbb{R}, i = 1, ..., n$, are the reward functions of the joint angles.

The complete joint displacements space is denoted by $\Delta Q = \Delta Q_1 \times \cdots \times \Delta Q_n$. The joint displacements are $\Delta q = \left[\Delta q_{i,t}^\top, ..., \Delta q_{n,t}^\top \right]^\top \in \Delta Q, \Delta q_{i,t} \in \Delta Q_i$. The policies of the inverse kinematics solution are

$$q_i : \mathcal{X} \to \Delta Q_i \in q(x).$$

The joint updates are

$$\sigma = \left[\sigma_1, \cdots, \sigma_n \right]^\top,$$

where $\sigma_i = q_i + \Delta q_i$.

In a fully cooperative stochastic game [25], the reward functions are the same for all the joint DOFs, i.e., $\rho_1 = \cdots = \rho_n$. For robot control, the reinforcement learning algorithm is modeled as the multi-agent case [18]. It uses different rewards for each joint angle. In the task-space control, it is a fully cooperative stochastic game. In our MARL, the inverse kinematics solution (9.22) and (9.24) are rewritten as

$$q_i^*(x) = \arg\min_{\sigma_i} \min_{\sigma_1, ..., \sigma_{i+1}, ..., \sigma_n} [\rho(x, \sigma) + \gamma V^*(x_{t+1})],$$

$$q_i^*(x) = \arg\min_{\sigma_i} \min_{\sigma_1, ..., \sigma_{i+1}, ..., \sigma_n} Q^*(x, \sigma).$$

The main problem of the MARL is the curse of dimensionality [34], which is caused by the exponential growth of the discrete state-action space. There are many algorithms that deal with the curse of dimensionality such as probabilistic inference for learning control [35], proximal policy optimization [36], deterministic policy gradients [37], and k-nearest neighbors [22], such as. These algorithms use different approximators in order to reduce the computational complexity.

The poor approximation may diverge and make the trajectories converge to wrong solutions [22].

For a redundant robot, each DOF needs one approximator; hence, the RL algorithm becomes more complex. This complexity implies more learning time and a rigorous hyper-parameters selection for both the approximator and the MARL algorithm.

In order to avoid function approximators in large state-action space, we use the actor-critic method for the MARL design. Our actor-critic method does not use any function approximator, since we only need discrete small displacements to update the forward kinematics. We do not need many joint displacements, because the kinematic learning algorithm updates the joint angles in each iteration by using the previous joint angle ΔQ_i. Even more, the proposed actor-critic method only has three hyper-parameters to choose: a learning rate for the critic and another one for the actor and the discount factor.

We use two value functions and the actor-critic method to obtain the inverse kinematics solution of each joint without any function approximator. The value function-based actor-critic method uses the state value function $V(\cdot)$ as the critic. The critic is updated by

$$V_{t+1}^{q(x)}(x_t) = V_t^{q(x)}(x_t) + \alpha_c \left(\rho(x, \sigma) + \gamma V_t^{q(x)}(x_{t+1}) - V_t^{q(x)}(x_t) \right), \tag{9.25}$$

where $\alpha_c > 0$ is the critic learning rate.

The action value function $Q(\cdot)$ is used to obtain the new control policy and the inverse kinematics. The actor is updated by

$$Q_{t+1}^{q(x)}(x_t, \sigma_t) = Q_t^{q(x)}(x_t, \sigma_t) + \alpha_a \delta_t, \tag{9.26}$$

where $\alpha_a > 0$ is the actor learning rate. The TD-error of (9.26) is

$$\delta_t = r_{t+1} + \gamma V_t^{q(x)}(x_{t+1}) - V_t^{q(x)}(x_t), \tag{9.27}$$

where $r_{t+1} = \rho(x, \sigma)$. The V-function uses the bootstrapping of the inverse kinematic solution $q(x)$, which is greedy and is obtained by

$$q_{i,t}^* = \arg\min_{\sigma_i} \ \min_{\sigma_1, \ldots, \sigma_{i+1}, \ldots, \sigma_n} \ Q_t^*(x_t, \sigma).$$

The calculation process of the inverse kinematics using MARL is given by Algorithm 9.1.

Remark 9.3   *When the robot is redundant, the learning space is huge. The MARL method uses the kinematic learning to avoid the curse of dimensionality. We use a small discrete action space for updating, which includes small robot joint positions. In task space, the input space to RL is composed of the end-effector pose and orientation. The proposed MARL can calculate the inverse kinematics online, and it needs less learning time in task space.*

Algorithm 9.1 MARL for inverse kinematics solution

1: **Input:** Initialize discount factor γ, actor-critic learning rates α_c, α_a, and desired position x_d.
2: **repeat** {for each episode}
3:     Initialize x_0
4:     **repeat** {for each step of episode $t = 0, 1, \ldots$ and joint $i = 1, \ldots, n$}
5:        Take displacement $\Delta q_{i,t}$, update $\sigma_{i,t} = q_{i,t} + \Delta q_{i,t}$, and observe r_{t+1}, x_{t+1}
6:        Update the critic by (9.25)
7:        Update the actor by (9.26)
8:        Choose $\Delta q_{i,t+1}$ from x_{t+1} greedily using any exploration technique.
9:        $x_t \leftarrow x_{t+1}, \sigma_{i,t} \leftarrow \sigma_{i,t+1}$
10:   **until** x_t is terminal.
11: **until** learning ends.

9.4 Simulations and experiments

We use a 4-DOF exoskeleton robot, shown in Figure A.1. The control objective of the robot is to move the end effector to a desired position. The dimension in task space is three, while the dimension in joint space is four. It is a redundant robot. We will compare our MARL with the following classical methods:

1) The Jacobian-based augmented method ($J^\dagger$ AJ);
2) The Jacobian-based singularity avoidance ($J^\dagger$ SA);
3) The Jacobian-based singular-value-decomposition ($J^\dagger$ SVD);
4) The multilayer neural networks (SLFNN).

The real-time environment is Simulink and MATLAB 2012. The communication is the CAN protocol, which enables the PC to communicate with the actuators. By differentiating (A.27) with respect to the joint angles, the analytic Jacobian $J(q)$ is obtained. The range of the analytic Jacobian is $\Re[J(q)] = 3$.

Example 9.1 *Simulations*

We first consider the singularity case. We set the initial joint positions as $q_0 = \left[0, \frac{\pi}{2}, 0, 0 \right]^T$. The pseudoinverse method can be applied. But $\det(J^T J) = 0$ for all joint positions, so (9.5) cannot be applied. We use SVD for the pseudoinverse Jacobian at the initial positions,

$$
J^\dagger = \begin{bmatrix} 0 & 0 & 0 \\ -1.8182 & 0 & 0 \\ 0 & 2.2727 & 0 \\ -0.9091 & 0 & 0 \end{bmatrix}.
$$

Because the last column is zero, $J^\dagger$ cannot give a correct mapping between the task space and the joint space. The augmented Jacobian method [38, 39] adds the angular velocity Jacobian (A.28) in the last column. The range of the augmented Jacobian is $\Re[J_a(q)] = 4$. Since the Jacobian is singular at the initial pose, we use the SVD method. The inverse of the Jacobian is

$$
J_a^{-1} = \begin{bmatrix} 0 & 0 & 0 & 1 \\ -1.8182 & 0 & 0 & 0 \\ 0 & 2.2727 & 0 & 0 \\ -0.9091 & 0 & 0 & 0 \end{bmatrix}.
$$

To avoid the singularity, we use the penalization constant $\xi = 0.1$ in (9.9).

The training set for the SLFNN is obtained using the homogeneous discretization of the joint space and the forward kinematics. The discretion area of the workspace for the 4-DOF exoskeleton is the ranges of the human shoulder and elbow movement,

$$
\frac{7\pi}{18} \le q_1 \le \frac{17\pi}{36}, \quad -\frac{11\pi}{36} \le q_2 \le \pi,
$$
$$
0 \le q_3 \le \frac{2\pi}{3}, \, 0 \le q_4 \le \frac{29\pi}{36}.
$$

We have 642,096 data points, and 70% of the available data are for training. There are 100 neurons at the hidden layer. The learning parameters of the MARL are $\alpha_a = 0.1$, $\alpha_c = 0.3$, and $\gamma = 0.9$. We use the ε-greedy exploration strategy (see equation (B.27)) with $\varepsilon = 0.01$. The MARL runs for 1,000 episodes, with 1,000 steps per episode. The desired position is $x_d = [-0.24, -0.196, 0.207]^T$. The reward is designed as the quadratic error between the desired reference and the robot pose:

$$
\rho(x, \sigma) = \|x_d - x\|^2 = \|x_d - f(\sigma)\|^2.
$$

We use the joint and task space PID control laws whose gains are given in Table 9.1 where I is an identity matrix with appropriate dimension.

The results at the Z axis are given in Figure 9.4(a)-9.4(c). The pseudoinverse methods using the SVD and the AJ show that they cannot achieve the control task, because the pseudoinverse matrix loses a degree of freedom. The SA method

Table 9.1 PID control gains

Gain	SVD	AJ	SA	SLFNN
K_p	$20I$	$30I$	$20I$	$9I$
K_d	$12I$	$20I$	$12I$	$4I$
K_i	$8I$	$14I$	$8I$	$2I$

Figure 9.4 Position tracking of simulations.

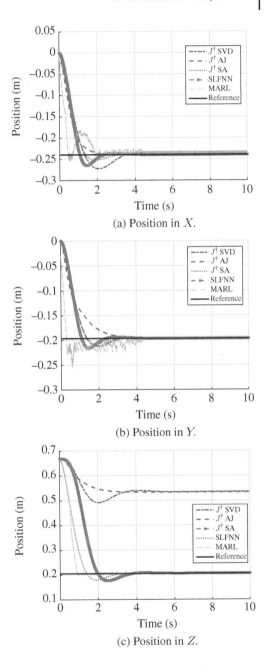

(a) Position in X.

(b) Position in Y.

(c) Position in Z.

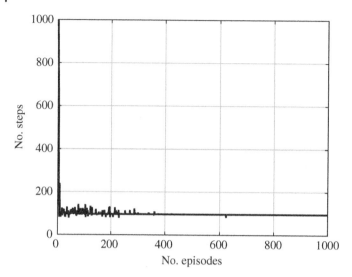

Figure 9.5 MARL Learning curve.

requires knowledge of the robot kinematics. The Jacobian-based methods produce high joint rates when the robot is at singular point, so the parameter ξ must be selected carefully.

The SLFNN requires off-line training, and it takes about 6 hours due to the complexity of the neural networks. The MARL algorithm converges to the desired reference faster than the other methods. The algorithm achieves the control task without controllability problems. The learning curve in Figure 9.5 shows that the algorithm converges rapidly, since the critic helps the actor to learn the optimal solution in less learning time. Figure 9.6 shows how our MARL minimizes the total cumulative reward in each learning episode, i.e., $\sum_{i=1}^{T} \rho_i(x, \sigma)$.

Example 9.2 *Experiments*

Now we compare our MARL with $J^{\dagger}$ SA and SLFNN. The desired pose is $x_d = [-0.086, 0.138, 0.62]^{T}$, and the initial joint position is the same as the singular case. We use the same learning parameters. The PID gains are modified as $K_p = 90I$, $K_d = 2I$, and $K_i = 5I$. In task space, $K_p = 180I$, $K_d = 4I$, and $K_i = 10I$. The comparison results are given in Figure 9.7.

We can see that $J^{\dagger}$ SA depends on the controller gains and the penalization constant, which determines the controller response. The MARL is a model-free algorithm and arrives at a sub-optimal solution. The advantages of the MARL kinematic learning method is that we do not need the robot dynamics and possible

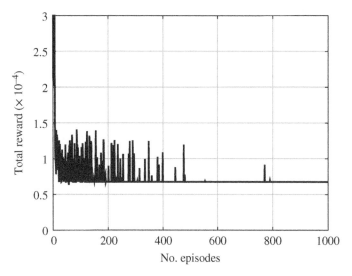

Figure 9.6 Total reward curve.

disturbances. MARL converges faster than the SLFNN and can be applied online. For more accurate responses we need to add new small displacements in the action space.

When the MARL converges, the inverse kinematic for the desired task is obtained. The inverse kinematics can be used for the repetitive tasks and to transform a task space control problem to a joint-space control problem in an easy way.

9.5 Conclusions

In this chapter the multi-agent reinforcement learning is applied for the redundant robot control in task space. This method does not need the inverse kinematics and Jacobian. This model-free controller avoids the task-space problems of redundant robots. The reinforcement learning algorithm uses two value functions to deal with the kinematic and controllability problems of robots. So the curse of dimensionality in RL and controllability problems for redundant robot control are overcome. The experiment results with a 4-DOF exoskeleton robot show our MARL approach is much better than the standard Jacobian-based and neural networks methods.

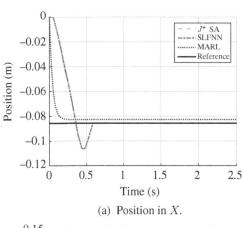

(a) Position in X.

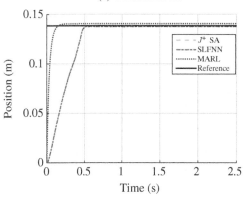

(b) Position in Y.

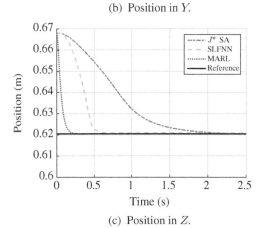

(c) Position in Z.

Figure 9.7 Position tracking of experiments.

References

1 D. Tuong and J. Peters, "Learning task-space tracking control with kernels," *IEEE/RSJ International Conference on Intelligent Robots and Systems*, 2011.

2 R. Patel and F. Shadpey, "Control of redundant manipulators: Theory and experiments," *Springer-Verlag: Berlin Heidelberg*, 2005.

3 Z. Szafrana, "Robust task space trajectory tracking control of robotic manipulators," *International Journal of Applied Mechanics and Engineering*, vol. 21, no. 3, pp. 547–568, 2016.

4 A. Perrusquía and W. Yu, "Task space human-robot interaction using angular velocity jacobian," *2019 International Symposium in Medical Robotics (ISMR)*, Atlanta, Georgia, pp. 1–7, 2019.

5 Y. Feng, W. Yao-nan, and Y. Yi-min, "Inverse kinematics solution for robot manipulator based on neural network under joint subspace," *International Journal of Computers Communications and Control*, vol. 7, no. 3, pp. 459–472, 2012.

6 M. Galicki, "Finite-time trajectory tracking control in task space of robotic manipulators," *Automatica*, vol. 67, pp. 165–170, 2016.

7 S. Ahmadi and M. Fateh, "Task-space asymptotic tracking control of robots using a direct adaptive taylor series controller,"*Journal of Vibration and Control*, vol. 24, no. 23, pp. 5570–5584, 2018.

8 L. Luya, W. Gruver, Q. Zhang, and Z. Yang, "Kinematic control of redundant robots and the motion optimizability measure.," *IEEE Transactions on Systems, man, and cybernetics-Part B: Cybernetics*, vol. 31, no. 1, 2001.

9 D. Axinte, X. Dong, D. Palmer, A. Rushworth, S. Guzman, and A. Olarra, "Miror—miniaturized robotic systems for holisticin-siturepair and maintenance works in restrained and hazardous environments," *IEEE/ASME Transactions on Mechatronics*, vol. 23, no. 2, pp. 978–981, 2018.

10 S. Atashzar, M. Tavakoli, and R. Patel, "A computational-model-based study of supervised haptics-enabled therapist-in-the-loop training for upper-limb poststroke robotic rehabilitation," *IEEE/ASME Transactions on Mechatronics*, vol. 23, no. 2, pp. 562–574, 2018.

11 C. Cheah and X. Li, "Singularity-robust task-space tracking control of robot," *IEEE International Conference on Robotics and Automation*, 2011.

12 B. Xian, M. de Queiroz, D. Dawson, and I. Walker, "Task-space tracking control of robots manipulators via quaternion feedback," *IEEE Transactions on Robotics and Automation*, vol. 20, no. 1, 2004.

13 D. Zhang and B. Wei, "On the development of learning control for robotic manipulators," *Robotics*, vol. 6, no. 23, 2017.

14 A. Csistzar, J. Eilers, and A. Verl, "On solving the inverse kinematics problem using neural networks," *Mechatronics and Machine Vision in Practice*, 2017.

15 M. Rolf and J. Steil, "Efficient exploratory learning of inverse kinematics on a bionic elephant trunk,"*IEEE Transactions on Neural Networks and Learning Systems*, vol. 25, no. 6, pp. 1147–1160, 2014.

16 B. Bcsi, D. Nguyen-Tuong, L. Csat, B. Schlkopf, and J. Peters, "Learning inverse kinematics with structured prediction," *IEEE/RSJ International Conference on Intelligent Robots and Systems*, 2011.

17 A. Duka, "Neural network based inverse kinematics solution for trajectory tracking of a robotic arm," *Procedia Technology, The 7th International Conference Interdisciplinarity in Engineering, INTER-ENG 2013*, 2014. Petru Maior University of Tirgu Mures, Romania.

18 B. Kiumarsi, K. G. Vamvoudakis, H. Modares, and F. L. Lewis, "Optimal and autonomous control using reinforcement learning: A survey,"*IEEE Transactions on Neural Networks and Learning Systems*, vol. 29, no. 6, 2018.

19 L. Buşoniu, R. Babûska, and B. De Schutter, "Multi-agent reinforcement learning: An overview," in *Innovations in Multi-Agent Systems and Applications - 1. Studies in Computational Intelligence* (D. In: Srinivasan and L. Jain, eds.), vol. 310, Springer: Berlin, Heidelberg, 2010.

20 A. Perrusquía, W. Yu, and A. Soria, "Position/force control of robots manipulators using reinforcement learning," *Industrial Robot: the international journal of robotics research and application*, vol. 46, no. 2, pp. 267–280, 2019.

21 R. Sutton and B. A, *Reinforcement Learning: An Introduction*. Cambridge, MA: MIT Press, 1998.

22 A. Perrusquiía and W. Yu, "Robust control under worst-case uncertainty for unknown nonlinear systems using modified reinforcement learning," *International Journal of Robust and Nonlinear Control*, 2020.

23 I. Grondman, L. Buşoniu, and R. Babûska, "Model learning actor-critic algorithms: performance evaluation in a motion control task," *51st IEEE Conference on Decision and Control (CDC)*, pp. 5272–5277, 2012b.

24 E. Theodorou, J. Buchli, and S. Schaal, "Reinforcement learning of motor skills in high dimensions: a path integral approach," *IEEE International Conference on Robotics an Automation (ICRA)*, 2010.

25 T. Tamei and T. Shibata, "Policy gradient learning of cooperative interaction with a robot using user's biological signals," *In International Conference on Neural Information Processing (ICONIP)*, 2009.

26 Y. Ansari and E. Falotico, "A multiagent reinforcement learning approach for inverse kinematics oh high dimensional manipulators with precision positioning," *6th IEEE RAS/EMBS International Conference on Biomedical Robotics and Biomechatronics (BioRob)*, 2016.

27 P. Hyatt, "Configuration estimation for accurate position control of large-scale soft robots," *IEEE/ASME Transactions on Mechatronics*, vol. 24, no. 1, pp. 88–99, 2019.

28 M. P. Deisenroth, G. Neumann, and J. Peters, "A survey on policy search for robotics," *Foundations and Trends in Robotics*, vol. 2, no. 1-2, pp. 1–142, 2011.

29 I. Grondman, M. Vaandrager, L. Buşoniu, R. Babûska, and E. Schuitema, "Actor-critic control with reference model learning," *Proc. of the 18th World Congress The International Federation of Automatic Control*, pp. 14723–14728, 2011.

30 I. Grondman, M. Vaandrager, L. Buşoniu, R. Babûska, and E. Schuitema, "Efficient model learning methods for actor-critic control,"*IEEE Transactions on Systems, man, and cybernetics. Part B: Cybernetics*, vol. 42, no. 3, 2012a.

31 S. Bitzer, M. Howard, and S. Vijayakumar, "Using dimensionality reduction to exploit constraints in reinforcement learning," *IEEE/RSJInternational Conference on Intelligent Robots and Systems (IROS)*, 2010.

32 F. Lewis, D. Vrable, and K. Vamvoudakis, "Reinforcement learning and feedback control using natural decision methods to desgin optimal adaptive controllers," *IEEE Control Systems Magazine*, 2012.

33 Y. Zhu, S. Li, J. Ma, and Y. Zheng, "Bipartite consensus in networks of agents with antagonistic interactions and quantization," *IEEE Transactions on Circuits and Systems II: Express Briefs*, vol. 65, no. 12, pp. 2012–2016, 2018.

34 A. Perrusquía, W. Yu, and A. Soria, "Large space dimension reinforcement learning for robot position/force discrete control," *2019 6th International Conference on Control, Decision and Information Technologies (CoDIT 2019)*, Paris, France, 2019.

35 M. Deisenroth and C. Rasmussen, "PILCO: A model-based and data-efficient approach to policy search," *in Proceedings of the 28th International Conference on Machine Learning*, 2011. Bellevue, Washington.

36 J. Schulman, F. Wolski, and O. Klimov, "Proximal policy optimization algorithms," *ArXiv*, 2017.

37 D. Silver, G. Lever, N. Hess, T. Degris, D. Wierstra, and M. Riedmiller, "Deterministic policy gradient algorithms," *Proceedings of the 31st International Conference on Machine Learning*, 2014. Beijing, China.

38 A. Perrusquía, W. Yu, A. Soria, and R. Lozano, "Stable admittance control without inverse kinematics," *20th IFAC World Congress (IFAC2017)*, 2017.

39 W. Yu and A. Perrusquía, "Simplified stable admittance control using end-effector orientations," *International Journal of Social Robotics*, 2019.

10

Robot $\mathcal{H}_2$ Neural Control Using Reinforcement Learning

10.1 Introduction

The aim of optimal and robust control is to find the controller that minimizes or maximizes a certain index or cost function according to a desired performance [1]. The most popular approach to optimal control is $\mathcal{H}_2$, such as the linear quadratic regulator (LQR) control. It uses the system dynamics to compute (off-line) the controller that minimizes/maximizes a cost function. To obtain the optimal controller online we can use the Hewer algorithm or Lyapunov recursion [2].

When the system dynamics is unknown, the classical $\mathcal{H}_2$ control cannot be applied directly, and other techniques are required [3]. The literature proposes adaptive dynamic programming (ADP) or reinforcement learning (RL) for unknown system dynamics [4–6]. The classical RL methods are designed in discrete time, and they need approximators, such as Gaussian kernels, linear parameterization, or neural networks [7–9], to deal with a large state-action space. These approximators need a long learning time due the exploration phase of the large input space [10–12].

In order to accelerate the learning time, some authors used the long short-term memory, such as eligibility traces [7], to visit states in previous steps. Other methods use model learning [13, 14] or reference-model learning [15], where the learned model serves as experience and exploits its knowledge for fast bootstrapping. They need accurate approximators to obtain a reliable solution of the optimal control problem. Neural networks is one of the most widely used approximators for RL methods and model-learning approaches. The main advantage of neural networks is that they can estimate and control complex systems by its feedback principle [16, 17]. However, their learnings need exploration terms, such as the persistent excitation (PE) signal, for good identification [11, 12, 18].

Another methodology to accelerate the learning time, which is not well established in current literature, is the use of recurrent neural networks (RNN) [19–21]. The stable dynamics are applied for the weights updating. This method is extended

Human-Robot Interaction Control Using Reinforcement Learning, First Edition. Wen Yu and Adolfo Perrusquía.
© 2022 The Institute of Electrical and Electronics Engineers, Inc. Published 2022 by John Wiley & Sons, Inc.

to the discrete-time neural networks by the gradient descent rule [22, 23] and the robust modifications [24] to guarantee the stability of the system identification. However, the neural identification is sensitive to modeling error and cannot guarantee optimal performance.

In this chapter, we discuss $\mathcal{H}_2$ neural control based on recurrent neural networks and reinforcement learning for robots, which has better performance than the standard LQR control and also accelerates the learning process. The controllers are designed in both discrete time and continuous time. The stability analysis and convergence of the neural identifier and the neural control are given.

10.2 $\mathcal{H}_2$ Neural Control Using Discrete-Time Reinforcement Learning

Control of discrete-time systems is becoming important in recent years since almost all of the control schemes are implemented on digital devices. Also machine learning techniques are developed in discrete time [1].

Consider the following discrete-time state-space nonlinear system:

$$x_{k+1} = f(x_k, u_k), \tag{10.1}$$

where $x_k \in \mathbb{R}^n$ is the state vector, $u_k \in \mathbb{R}^m$ is the control vector, and $f \in \mathbb{R}^n$ is the unknown nonlinear transition function.

We use the following discrete-time serial-parallel recurrent neural network [23] to model (10.1),

$$\hat{x}_{k+1} = A\hat{x}_k + W_{1,k}\sigma(W_{2,k}x_k) + U_k, \tag{10.2}$$

where $\hat{x}_k \in \mathbb{R}^n$ is the state of the RNN and

$$U_k = [u_1, u_2, \cdots, u_m, 0, \cdots, 0]^\top \in \mathbb{R}^n$$

is the control input. The matrix $A \in \mathbb{R}^{n \times n}$ is a stable Hurwitz matrix, which will be specified after, $W_{1,k} \in \mathbb{R}^{n \times r}$ is the output weights, $W_{2,k} \in \mathbb{R}^{l \times n}$ is the weight of the hidden layer, and $\sigma(\cdot) : \mathbb{R}^l \to \mathbb{R}^r$ is the activation function, where $\sigma_i(\cdot)$ can be any stationary, bounded, and monotone increasing function.

For nonlinear system modeling, we consider the following two cases:

1) We assume that the neural model (10.2), with fixed $W_{2,k}$, can exactly approximate the nonlinear system (10.1). According to the Stone-Weierstrass theorem [25], the nonlinear model (10.1) can be written as

$$x_{k+1} = Ax_k + W_1^*\sigma(W_2^*x_k) + U_k, \tag{10.3}$$

where W_1^* and W_2^* are optimal weight matrices.

In order to obtain the parameters update rule, we expand the term $\sigma(W_{2,k}x_k)$ of (10.2) with the following Taylor formula [22],

$$g(x) = \sum_{i=0}^{l-1} \frac{1}{i!} \left[(x_1 - x_1^0)\frac{\partial}{\partial x_1} + (x_2 - x_2^0)\frac{\partial}{\partial x_2} \right]_0^i g(x^0) + \varepsilon_g$$

where $x = [x_1, x_2]^\mathsf{T}$, $x^0 = [x_1^0, x_2^0]^\mathsf{T}$, and ε_g is the remainder of the Taylor formula. Let x_1 and x_2 correspond to $W_{1,k}$ and $W_{2,k}x_k$, respectively, x_1^0 and x_2^0 correspond to W_1^* and $W_2^*x_k$, respectively, and $l = 2$. The identification error is defined as

$$\tilde{x}_k = \hat{x}_k - x_k. \tag{10.4}$$

From (10.3) and (10.2),

$$\tilde{x}_{k+1} = A\tilde{x}_k + \widetilde{W}_{1,k}\sigma(W_{2,k}x_k) + W_1\sigma'\widetilde{W}_{2,k}x_k + \varepsilon_W, \tag{10.5}$$

where $\widetilde{W}_{1,k} = W_{1,k} - W_1^*$, $\widetilde{W}_{2,k} = W_{2,k} - W_2^*$, and ε_W is a second-order approximation error of the Taylor series. When $W_{2,k}$ is fixed, $\varepsilon_W = 0$, so

$$\tilde{x}_{k+1} = A\tilde{x}_k + \widetilde{W}_{1,k}\sigma(W_2 x_k). \tag{10.6}$$

2) Generally, the neural network (10.2) cannot fully describe the nonlinear system (10.1). The nonlinear system (10.1) can be expressed as

$$x_{k+1} = Ax_k + W_{1,k}\sigma(W_{2,k}x_k) + U_k + \zeta_k, \tag{10.7}$$

where $\zeta_k = -\widetilde{W}_{1,k}\sigma(W_{2,k}x_k) - W_{1,k}\sigma'\widetilde{W}_{2,k}x_k - \varepsilon_f + \varepsilon_W$, ε_f is the modeling error defined as

$$\varepsilon_f = f(x_k, u_k) - Ax_k - W_1^*\sigma(W_2^*x_k) - U_k. \tag{10.8}$$

The dynamics of the identification error (10.4) is

$$\tilde{x}_{k+1} = A\tilde{x}_k + \widetilde{W}_{1,k}\sigma(W_{2,k}\hat{x}_k) + W_{1,k}\sigma'\widetilde{W}_{2,k}x_k + \xi_k, \tag{10.9}$$

where $\xi_k = \varepsilon_W - \varepsilon_f$.

We use the following assumptions:

Assumption 10.1 There exists a constant $\beta \geq 1$ such that

$$\|x_{k+1}\| \geq \frac{1}{\beta}\|x_k\|. \tag{10.10}$$

This condition in fact is a dead zone [24]. We can select β big enough so that the dead-zone is almost disperse.

Assumption 10.2 The unmodeled dynamic ξ_k in (10.7) is bounded:

$$\|\xi_k\| \leq \bar{\xi}. \tag{10.11}$$

If the matrix A in (10.2) is selected as a stable matrix, then we have that the following property holds:

P10.1: If the eigenvalues of matrix A are in the interval $-\frac{1}{\beta} < \lambda(A) < 0$, then for any matrix $Q^i = Q^{i^T} > 0$, there exists a unique solution $P^i = P^{i^T} > 0$ for the following Lyapunov equation:

$$A^T P^i A - P^i + Q^i + \left(\frac{1}{\beta}I + A\right)^T P^i \left(\frac{1}{\beta}I + A\right) = 0. \tag{10.12}$$

The solution of the Lyapunov equation is

$$\left\{ I - A^T \bigotimes A^T - \left[\frac{1}{\beta}I + A\right]^T \bigotimes \left[\frac{1}{\beta}I + A\right]^T \right\} \operatorname{vec}(P^i) = \operatorname{vec}(Q^i).$$

where $\bigotimes$ is the Kronecker product, and $\operatorname{vec}(\cdot)$ is the matrix stretch.

Property P10.1 is used to prove the stability of the neural identifier under the condition $-\frac{1}{\beta} < \lambda(A) < 0$. We will design an online $\mathcal{H}_2$ control using two different models: recurrent neural networks (RNN) and reinforcement learning (RL). Both approaches are designed for the above two cases: with exact model matching and with modeling error.

We first design the critic control to follow the desired reference $x_d \in \mathbb{R}^n$, which is the solution of the following model:

$$x_{k+1}^d = \varphi(x_k^d). \tag{10.13}$$

The tracking error is

$$e_k = x_k - x_k^d. \tag{10.14}$$

The nonlinear system (10.1) is represented by the neural network (10.2) with the unmodeled dynamic ζ_k:

$$x_{k+1} = Ax_k + W_{1,k}\sigma(W_{2,k}x_k) + U_k + \zeta_k, \tag{10.15}$$

where $\zeta_k = -\widetilde{W}_{1,k}\sigma(W_{2,k}x_k) - W_{1,k}\sigma'\widetilde{W}_{2,k}x_k - \xi_k$. The closed-loop error dynamics between (10.15) and (10.13) is

$$e_{k+1} = Ax_k + W_{1,k}\sigma(W_{2,k}x_k) + U_k + \zeta_k - \varphi(x_k^d). \tag{10.16}$$

We design the following feedback-feedforward control U_k for the unknown nonlinear system (10.1):

$$U_k = U_1 + U_2$$
$$= \varphi(x_k^d) - Ax_k^d - W_{1,k}\sigma(W_{2,k}x_k) + U_2, \tag{10.17}$$

where U_1 is the neural feedforward control, which uses the neural model (10.15), and U_2 is the $\mathcal{H}_2$ feedback control.

The closed-loop error dynamics (10.16) is reduced to

$$e_{k+1} = Ae_k + U_2 + \zeta_k. \tag{10.18}$$

To design the feedback control U_2, the following property is used.

P10.2: There exists a strictly positive definite matrix Q^c such that the discrete algebraic Riccati equation (DARE)

$$A^\top P^c A + Q^c - A^\top P^c (R^c + P^c)^{-1} P^c A - P^c = 0 \tag{10.19}$$

has a positive solution $P^c = P^{c^\top} > 0$, $R^c = R^{c^\top} > 0$.

According to the property P10.2, the feedback control U_2 has the following form

$$U_2 = -Ke_k, \quad K = -(R^c + P^c)^{-1} P^c A, \tag{10.20}$$

where $R^c = R^{c^\top} > 0$, P_j^c is the solution of the DARE (10.19). The solution of (10.19) can be obtained iteratively using Lyapunov recursions [26],

$$P_{k+1}^c = (A - K_k)^\top P_k^c (A - K_k) + Q^c + K_k^\top R^c K_k. \tag{10.21}$$

Since $(A - K_k)$ is stable, (10.21) converges to the DARE (10.19) solution with any initial value P_0^c.

The following theorem gives the asymptotic stability of the neural $\mathcal{H}_2$ control if the neural model (10.2) can exactly approximate the nonlinear system (10.1).

Theorem 10.1 *If the unknown nonlinear dynamics (10.1) can be approximated by the neural model (10.2) exactly with fixed W_2, the weight of the neural model are adjusted by*

$$\widetilde{W}_{1,k+1} = \widetilde{W}_{1,k} - \eta P^i A \widetilde{x}_k \sigma^\top \tag{10.22}$$

with

$$\eta = \begin{cases} \eta_0 \dfrac{1}{1 + \|P^i A \sigma\|^2} & \text{if } \|\widetilde{x}_{k+1}\| \geq \frac{1}{\beta} \|\widetilde{x}_k\| \\ 0 & \text{otherwise,} \end{cases} \tag{10.23}$$

where $\eta_0 \in (0, 1]$. Under Assumption 10.1, the tracking error of the neural $\mathcal{H}_2$ control (10.17) is globally asymptotically stable.

Proof: Consider the Lyapunov function

$$V_k = \widetilde{x}_k^\top P^i \widetilde{x}_k + e_k^\top P^c e_k + \frac{1}{\eta} \text{tr}(\widetilde{W}_{1,k}^\top \widetilde{W}_{1,k}), \tag{10.24}$$

where $\widetilde{x}_k$ is the neural modeling error, and e_k is the tracking error e_k of the $\mathcal{H}_2$ control. The time difference of the Lyapunov equation $\Delta V_k = V_{k+1} - V_k$ is

$$\begin{aligned} \Delta V_k &= \widetilde{x}_{k+1}^\top P^i \widetilde{x}_{k+1} + e_{k+1}^\top P^c e_{k+1} - \widetilde{x}_k^\top P^i \widetilde{x}_k - e_k^\top P^c e_k \\ &\quad + \frac{1}{\eta} \left(\text{tr}(\widetilde{W}_{1,k+1}^\top \widetilde{W}_{1,k+1}) \right) - \frac{1}{\eta} \left(\text{tr}(\widetilde{W}_{1,k}^\top \widetilde{W}_{1,k}) \right). \end{aligned}$$

ΔV_k includes the identification part $\Delta V_{1,k}$ and the control part $\Delta V_{2,k}$, which can be written as

$$\Delta V_{1,k} = \Delta V_k - \Delta V_{2,k},$$
$$\Delta V_{2,k} = e_{k+1}^{\mathsf{T}} P^c e_{k+1} - e_k^{\mathsf{T}} P^c e_k.$$

Substituting the identification error dynamics (10.5) and the updating law (10.22) in $\Delta V_{1,k}$,

$$\Delta V_{1,k} = \tilde{x}_k^{\mathsf{T}}(A^{\mathsf{T}} P^i A - P^i)\tilde{x}_k + \sigma^{\mathsf{T}} \widetilde{W}_{1,k}^{\mathsf{T}} P \widetilde{W}_{1,k} \sigma$$
$$+ 2\tilde{x}^{\mathsf{T}} A^{\mathsf{T}} P^i \widetilde{W}_{1,k} \sigma - 2\mathrm{tr}(\widetilde{W}_{1,k}^{\mathsf{T}} Z) + \eta\,\mathrm{tr}(Z^{\mathsf{T}} Z),$$

where $Z = P^i A \tilde{x}_k \sigma^{\mathsf{T}}$. The above expression is simplified by the Minkowski inequality:

$$\Delta V_{1,k} \le \tilde{x}_k^{\mathsf{T}}(A^{\mathsf{T}} P^i A - P^i)\tilde{x}_k + \|P^i\| \|\tilde{x}_{k+1} - A\tilde{x}_k\|^2 + \|P^i A \tilde{x}_k\|^2. \tag{10.25}$$

Using Assumption 10.1, the second term of (10.25) can be written as

$$\|P^i\| \|\tilde{x}_{k+1} - A\tilde{x}_k\|^2 \le \|P^i\| (\|\tilde{x}_{k+1}\|^2 + \|A\tilde{x}_k\|^2)$$
$$\le \|P^i\| \left(\frac{1}{\beta^2} + \|A\|^2 \right) \|\tilde{x}_k\|^2 \le \|P^i\| \|\tfrac{1}{\beta} I + A\|^2 \|\tilde{x}_k\|^2$$
$$= \tilde{x}_k^{\mathsf{T}} \left(\frac{1}{\beta} I + A \right)^{\mathsf{T}} P^i \left(\frac{1}{\beta} I + A \right) \tilde{x}_k.$$

From property P10.1,

$$\Delta V_{1,k} \le - \left[\lambda_{\min}(Q^i) - \eta_0 \frac{\|P^i A \sigma\|^2}{1 + \|P^i A \sigma\|^2} \right] \|\tilde{x}_k\|^2 \le -\alpha_0 \|\tilde{x}_k\|^2,$$

where $\alpha_0 = \lambda_{\min}(Q^i) - \eta_0 \frac{\kappa}{1+\kappa^2} > 0$, and $\kappa = \max_k \|P^i A \sigma\|$. From Barbalat's lemma, $\tilde{x}_k$ converges to zero and we can conclude asymptotic stability of the neural network identification.

In an exact approximation case, the error dynamics of the tracking control (10.18) is

$$e_{k+1} = Ae_k + U_2.$$

Using the discrete algebraic Riccati equation (10.19) and (10.20), the control part $\Delta V_{2,k}$ is

$$\Delta V_{2,k} = e_{k+1}^{\mathsf{T}} P^c e_{k+1} - e_k^{\mathsf{T}} P^c e_k \pm e_k^{\mathsf{T}} Q^c e_k \pm U_2^{\mathsf{T}} R^c U_2$$
$$= e_k^{\mathsf{T}}(A^{\mathsf{T}} P^c A - P^c \pm Q^c)e_k + U_2^{\mathsf{T}}(R^c + P^c)U_2 + 2e_k^{\mathsf{T}} A^{\mathsf{T}} P^c U_2$$
$$= -e_k^{\mathsf{T}}(Q^c + K^{\mathsf{T}} R^c K)e_k$$
$$= -e_k^{\mathsf{T}} \overline{Q} e_k \le 0,$$

where $\overline{Q} = Q^c + K^{\mathsf{T}} R^c K$. From Barbalat's lemma, e_k is asymptotically stable. ∎

Remark 10.1 *We fix the hidden weight W_2 to simplify the neural modeling process. W_2 can be updated by the backpropagation method as in [22]. On the other hand, if the hidden weight W_2 is chosen randomly, the neural networks have good approximation ability. Here, the elements of W_2 are chosen randomly in $(0,1)$ and then are fixed.*

In the general case, the neural model (10.2) cannot exactly approximate the nonlinear system (10.1) as (10.7). The following theorem gives the stability results of the neural H_2 control.

Theorem 10.2 *The nonlinear system (10.1) is controlled by the neural H_2 control (10.17). If the weights of the recurrent neural network (10.2) are adjusted by*

$$\widetilde{W}_{1,k+1} = \widetilde{W}_{1,k} - \eta \left(P^i A \widetilde{x}_k + W_1 \sigma' \widetilde{W}_{2,k} x_k \right) \sigma^\top$$
$$\widetilde{W}_{2,k+1} = \widetilde{W}_{2,k} - \eta \sigma'^\top W_{1,k}^\top P^i A \widetilde{x}_k x_k^\top, \tag{10.26}$$

where η satisfies

$$\eta = \begin{cases} \dfrac{\eta_0}{1+\|P^i A \sigma\|^2 + \|\sigma'^\top W_{1,k}^\top P^i A x_k\|^2} & \text{if } \|\widetilde{x}_{k+1}\| \geq \frac{1}{\beta}\|\widetilde{x}_k\| \\[4pt] 0 & \text{otherwise,} \end{cases} \tag{10.27}$$

with $\eta_0 \in (0,1]$. Under Assumption 10.1 and Assumption 10.2, the tracking error e_k is stable and converges into a small bounded set.

Proof: Because $W_{2,k}$ is not constant as in Theorem 10.1, we change the Lyapunov function as

$$V_k = \widetilde{x}_k^\top P^i \widetilde{x}_k + e_k^\top P^c e_k + \frac{1}{\eta} \left[\text{tr}(\widetilde{W}_{1,k}^\top \widetilde{W}_{1,k}) + \text{tr}(\widetilde{W}_{2,k}^\top \widetilde{W}_{2,k}) \right]. \tag{10.28}$$

Similar as the proof of Theorem 10.1, ΔV_k is also divided in $\Delta V_{1,k}$ and $\Delta V_{2,k}$. Because there is modeling error ζ_k, the identification error dynamics is changed from (10.6) to (10.9),

$$\begin{aligned} \Delta V_{1,k} = {}& \widetilde{x}_k^\top (A^\top P^i A - P^i) \widetilde{x}_k + \sigma^\top \widetilde{W}_{1,k}^\top P^i \widetilde{W}_{1,k} \sigma \\ &+ 2 \widetilde{x}^\top A^\top P^i (\widetilde{W}_{1,k} \sigma + W_{1,k} \sigma' \widetilde{W}_{2,t} x_k + \zeta_k) \\ &+ 2\sigma^\top \widetilde{W}_{1,k}^\top P^i (W_{1,k} \sigma' \widetilde{W}_2 x_k + \zeta_k) + \varepsilon_W^\top P^i \zeta_k \\ &+ x_k^\top \widetilde{W}_{2,k}^\top \sigma'^\top W_{1,k}^\top P^i W_{1,k} \sigma' \widetilde{W}_{2,k} x_k + \eta \text{tr}(Z_1^\top Z_1) \\ &+ 2\widetilde{x}^\top \widetilde{W}_{2,k}^\top \sigma'^\top W_{1,k}^\top P^i \zeta_k - 2\text{tr}(\widetilde{W}_{1,k}^\top Z_1) \\ &+ 2\text{tr}(\widetilde{W}_{2,k}^\top Z_2) + \eta \text{tr}(Z_2^\top Z_2), \end{aligned}$$

where $Z_1 = (P^i A \widetilde{x}_k + P^i W_1 \sigma' \widetilde{W}_{2,k} x_k) \sigma^\top$ and $Z_2 = \sigma'^\top W_{1,k}^\top P^i A \widetilde{x}_k x_k^\top$. Finally,

$$\Delta V_{1,k} \le -\lambda_{\min}(Q^i) \|\widetilde{x}_k\|^2 + \frac{1}{\beta^2} \|\widetilde{x}_k\|^2 + \eta_0 \frac{\|P A \sigma\|^2 + \|\sigma'^\top W_{1,k}^\top P^i A x_k\|^2}{1 + \|P^i A \sigma\|^2 + \|\sigma'^\top W_{1,k}^\top P^i A x_k\|^2} \|\widetilde{x}_k\|^2$$
$$+ \left[\lambda_{\max}^2(P^i) - 2\lambda_{\min}(P^i)\right] \|\zeta_k\|^2$$
$$\le -\alpha_1 \|\widetilde{x}_k\|^2 + \chi \|\zeta_k\|^2,$$

where

$$\alpha_1 = \lambda_{\min}(Q^i) - \frac{1}{\beta^2} - \frac{\eta_0 \kappa}{1 + \kappa},$$
$$\chi = \lambda_{\max}^2(P^i) - 2\lambda_{\min}(P^i),$$
$$\kappa = \max_k (\|P^i A \sigma\|^2 + \|\sigma'^\top W_{1,k}^\top P^i A x_k\|^2).$$

Because

$$\lambda_{\min}(Q^i) - \frac{1}{\beta^2} - \eta_0 \frac{\kappa}{1 + \kappa} > \lambda_{\min}(Q^i) - \frac{1}{\beta^2} - \eta_0 > \lambda_{\min}(Q^i) - \frac{1}{\beta^2} - 1,$$

we can select big enough β such that P^i and Q^i are positive solutions of (10.12) and $\alpha_1 > 0$, $\chi > 0$. From the the input to state stability theory, when $\|\zeta_k\|^2$ is bounded by $\overline{\xi}$ (Assumption 10.2), the modeling error converges to

$$\|\widetilde{x}_k\| \to \sqrt{\frac{\lambda_{\max}(\chi_1)}{\lambda_{\min}(\alpha_1)}} \overline{\xi}.$$

The control part $\Delta V_{2,k}$ is

$$\Delta V_{2,k} = e_{k+1}^\top P^c e_{k+1} - e_k^\top P^c e_k \pm e_k^\top Q^c e_k \pm U_2^\top R^c U_2$$
$$= e_k^\top (A^\top P^c A - P^c \pm Q^c) e_k \pm U_2^\top (R^c + P^c) U_2$$
$$\quad + 2 e_k^\top A^\top P^c (U_2 + \zeta_k) + 2 U_2^\top P^c \zeta_k + \zeta_k^\top P^c \zeta_k$$
$$= -e_k^\top (Q^c + K^\top R^c K) e_k + \zeta_k^\top P^c \zeta_k - 2 e_k^\top K^\top P^c \zeta_k$$
$$\le -e_k^\top (Q^c + K^\top (R^c - I) K) e_k + \zeta_k^\top (P^c + P^{c^2}) \zeta_k$$
$$\le -e_k^\top \overline{Q} e_k + \zeta_k^\top \Omega \zeta_k,$$

where $\Omega = P^c + P^{c^2} > 0$, and $\overline{Q} = Q^c + K^\top (R^c - I) K > 0$. From the input to state stability theory, when $\|\zeta_k\|^2$ is bounded by $\overline{\xi}$ (Assumption 10.2), the tracking error converges to

$$\|e_k\| \to \sqrt{\frac{\lambda_{\max}(\Omega)}{\lambda_{\min}(Q^c)}} \overline{\xi}.$$

∎

Remark 10.2 *The dead zones η_k in (10.23) and (10.27) could produce a small chattering effect at the weights updates. The size of the dead zones is $\frac{1}{\beta}$. From Assumption 10.1, it is not difficult to choose a big enough β such that the dead zones become very small. This small chattering does not affect the closed-loop identification and control.*

The RNN $\mathcal{H}_2$ control for the general case is summarized in Algorithm 10.1.

Algorithm 10.1 RNN $\mathcal{H}_2$ control.

1: **print** Matrices A, Q^i, Q^c, R^c, $W_{1,0}$, and $W_{2,0}$. Scalars β and η_0. Activation functions $\sigma(\cdot)$. Desired reference $\varphi(x^d)$.
2: Compute the kernel matrix P^i using (10.12)
3: Compute the kernel matrix P^c using (10.19)
4: **for** each step $k = 0, 1, \ldots$ **do**
5:     Measure state x_k
6:     Compute the RNN identifier using (10.2).
7:     Calculate the identification error $\tilde{x}_k$ using (10.4)
8:     Calculate the tracking error e_k using (10.63)
9:     Update weights $W_{1,k+1}$ and $W_{2,k+1}$ using (10.26) and (10.27)
10:    Apply the feedback-feedforward control U_1 and U_2 using (10.17) and (10.20)
11: **end for**

The $\mathcal{H}_2$ controller U_2 does not need any information of the real system (10.1). It assumes that all nonlinear dynamics are compensated by the neural controller U_1. The neural control U_1 is sensitive to the neural modeling error $\tilde{x}_k$. U_1 may have a big tracking control error with even small modeling error ξ_k.

We will use reinforcement learning for the $\mathcal{H}_2$ control. Since U_2 uses the whole tracking error e_k to learn the control errors caused by both U_1 and U_2, the reinforcement learning control can overcome the problem in the modeling process. To overcome the problem caused by the modeling error, we use reinforcement learning to avoid knowledge of the tracking error dynamics and use only measurements of the error state e_k to compute the feedback controller U_2.

For an $\mathcal{H}_2$ control, we define the following discounted cost function [5] subject to the error dynamics (10.18):

$$S_k = \gamma_1 S_{k+1} + e_k^\top Q^c e_k + U_2^\top R^c U_2, \tag{10.29}$$

where $S_k = \sum_{i=k}^{\infty} \gamma_1^{i-k} \left(e_i^\top Q^c e_i + U_2^\top R^c U_2 \right)$, and $\gamma_1[0, 1)$ is a discounted factor that guarantees the convergence of S_k.

Consider the following neural network approximator:

$$\hat{S}_k = \phi^\top(e_k)\theta_k = \phi_k^\top \theta_k, \tag{10.30}$$

where $\theta_k \in \mathbb{R}^p$ is a weight vector, and $\phi(\cdot) : \mathbb{R}^n \to \mathbb{R}^p$ are the activation functions with p neurons at the hidden layer. The value function S_k can be rewritten as

$$S_k = \phi_k^\top \theta^* + \varepsilon_k, \tag{10.31}$$

where θ^* is the unknown optimal weight value, and $\varepsilon_k = \varepsilon(e_k)$ is the neural approximation error. Substituting (10.31) into (10.29) yields

$$\phi_k^T \theta^* + \varepsilon_k = e_k^T Q^c e_k + U_2^T R^c U_2 + \gamma_1(\phi_{k+1}^T \theta^* + \varepsilon_{k+1}),$$
$$\varepsilon(e_k) - \gamma_1 \varepsilon(e_{k+1}) = e_k^T Q^c e_k + U_2^T R^c U_2 + (\gamma_1 \phi_{k+1}^T - \phi_k^T)\theta^*.$$

We define

$$H(e_k, \theta^*) = r_{k+1} + (\phi_{k+1}^T - \phi_k^T)\theta^* = v_k, \tag{10.32}$$

where $v_k = \varepsilon(e_k) - \gamma_1 \varepsilon(e_{k+1})$ is the residual error of the NN approximator, which is equivalent to the discrete-time Hamiltonian, and

$$r_{k+1} = e_k^T Q^c e_k + U_2^T R^c U_2$$

is the immediate reward or utility function. The Hamiltonian can be approximated as

$$\hat{H}(e_k, \theta_k) = r_{k+1} + (\gamma_1 \phi_{k+1}^T - \phi_k^T)\theta_k = \delta_k. \tag{10.33}$$

The temporal difference error δ_k can be also be written as

$$\delta_k = (\gamma_1 \phi_{k+1}^T - \phi_k^T)\tilde{\theta}_k + v_k, \tag{10.34}$$

where $\tilde{\theta}_k = \theta_k - \theta^*$. The object of the reinforcement learning is to minimize the residual error δ_k. We define the objective function as the squared temporal difference error,

$$E = \frac{1}{2}\delta_k^2.$$

The weights of the neural approximator are updated as the normalized gradient descent algorithm

$$\theta_{k+1} = \theta_k - \alpha \frac{\partial E}{\partial \theta_k} = \theta_k - \alpha \delta_k \frac{q_k}{(q_k^T q_k + 1)^2}, \tag{10.35}$$

where $\alpha \in (0, 1]$ is the learning rate, $q_k^T = (\gamma_1 \phi_{k+1}^T - \phi_k^T)$. The update rule can be rewritten as

$$\tilde{\theta}_{k+1} = \tilde{\theta}_k - \alpha \frac{q_k q_k^T}{(q_k^T q_k + 1)^2}\tilde{\theta}_k - \alpha \frac{q_k}{(q_k^T q_k + 1)^2} v_k. \tag{10.36}$$

The $\mathcal{H}_2$ control has the following form,

$$U_2 = -\frac{\gamma_1}{2}R^{c^{-1}}\nabla\phi^T(e_{k+1})\theta_k, \tag{10.37}$$

where $\nabla = \partial/\partial e_{k+1}$. Substituting the $\mathcal{H}_2$ control (10.37) into the Hamiltonian yields,

$$\delta_k = e_k^T Q^c e_k - \frac{\gamma_1^2}{4}\theta_k^T \nabla\phi_{k+1}R^{c^{-1}}\nabla\phi_{k+1}^T\theta_k + (\phi_{k+1}^T - \phi_k^T)\theta_k.$$

This normalized gradient descent guarantees the boundedness of the weights update [11]. The convergence of the reinforcement learning corresponds to how $\hat{H}(\theta_k)$ or δ_k converges to $H(\theta^*)$ or v_k. This means the convergence of the parameters of neural approximators define the property of the $\mathcal{H}_2$ control based on reinforcement learning.

Now we analyze the convergence of the $\mathcal{H}_2$ control, or $\theta_k \to \theta^*$. We first define the persistently exciting (PE) condition.

Definition 10.1 *The learning input $q/(q^\top q + 1)$ is called persistently exciting (PE) in T steps if there exist constants $\beta_1, \beta_2 > 0$ such that*

$$\beta_1 I \leq S_1 = \sum_{j=k+1}^{k+T} \frac{q_j q_j^\top}{(q_j^\top q_j + 1)^2} \leq \beta_2 I. \tag{10.38}$$

Theorem 10.3 *If the learning input $q_k/(q_k^\top q + 1)$ in (10.35) is persistently exciting, the neural approximator error $\widetilde{\theta}_k = \theta_k - \theta^*$ and the neural approximator $F(\theta) = \phi_k^\top \theta_k$ are bounded as follows:*

$$\begin{aligned} \|\theta - \theta^*\| &\leq \frac{1+\gamma_2}{1-\gamma_2} \overline{v}, \\ \|F(\theta_k) - F(\theta^*)\| &\leq \frac{\gamma_2(1+\gamma_1)}{\gamma_2(1-\gamma_2)} \overline{v}, \end{aligned} \tag{10.39}$$

where γ_1 and γ_2 are contraction factors, and $\overline{v}$ is the upper bound of the residual error v_k.

Proof: Define the dynamic programming operator $M_1 : \mathcal{L} \to \mathcal{L}$ as in [9]:

$$M_1(S) = \min_u \left(r_{k+1} + \gamma S_{k+1} \right), \tag{10.40}$$

which is a contraction, that is, for all $S_1, S_2 \in \mathcal{L}$,

$$\|M_1(S_1) - M_1(S_2)\| \leq \gamma_1 \|S_1 - S_2\|. \tag{10.41}$$

Because M_1 is a contraction, it has a unique fixed point S^* that satisfies $S^* = M_1(S^*)$. We consider the value function approximation (10.30) and its dynamic programming operator $M_2 = F^\dagger M_1(F(\theta)) : \mathbb{R}^p \to \mathbb{R}^p$, where $F^\dagger$ is a pseudoinverse projection of F:

$$\|F^\dagger(S_1) - F^\dagger(S_2)\| \leq \|S_1 - S_2\|. \tag{10.42}$$

This implies that the dynamic programming operator M_2 is also a contraction, with contraction factor $\gamma_2 \in [\gamma_1, 1)$, i.e., for all $\theta_1, \theta_2 \in \mathbb{R}^p$,

$$\|M_2(\theta_1) - M_2(\theta_2)\| \leq \gamma_2 \|\theta_1 - \theta_2\|. \tag{10.43}$$

Define the residual approximation error as

$$\|v_k\| = \|S - F(\theta^*)\| \tag{10.44}$$

and consider an upper bound of the residual error $\bar{v} = \|v_k\| + \mu$ for some $\mu > 0$ that satisfies $\|S - F(\theta_x)\| \leq \bar{v}$ with $\theta_x \in \mathbb{R}^p$:

$$
\begin{aligned}
\|\theta_x - M_2(\theta_x)\| &\doteq \|F^\dagger F(\theta_x) - F^\dagger M_1(F(\theta_x))\| \\
&\leq \|F(\theta_x) - M_1(F(\theta_x))\| \\
&\leq \|F(\theta_x) - S\| + \|S - M_1(F(\theta_x))\| \\
&= (1 + \gamma_1)\bar{v}.
\end{aligned}
$$

Then the parameters error is bounded as follows:

$$
\begin{aligned}
\|\theta - \theta^*\| &= \|\theta - M_2(\theta_x)\| + \|M_2(\theta) - \theta^*\| \\
&< (1 + \gamma_1)\bar{v} + \gamma_2\|\theta - \theta^*\| < \frac{1+\gamma_1}{1-\gamma_2\bar{v}}.
\end{aligned}
$$

Since μ can be arbitrarily small, $\bar{v} \geq \|v_k\|$, and we conclude that the parameters error is bounded:

$$\|\tilde{\theta}\| < \frac{1 + \gamma_1}{1 - \gamma_2}\bar{v}.$$

Then the approximator is bounded as follows:

$$
\begin{aligned}
\|M_2(\theta) - M_2(\theta^*)\| &\leq \|F^\dagger M_1(F(\theta)) - F^\dagger M_1(F(\theta^*))\| \\
&\leq \|M_1(F(\theta)) - M_1(F(\theta^*))\| \\
&\leq \gamma_1\|F(\theta) - F(\theta^*)\| \leq \gamma_2\|\tilde{\theta}\|.
\end{aligned}
$$

Finally,

$$\|F(\theta) - F(\theta^*)\| \leq \frac{\gamma_2}{\gamma_1}\|\tilde{\theta}\| \leq \frac{\gamma_2(1 + \gamma_1)}{\gamma_1(1 - \gamma_2)}\bar{v}.$$

When $\gamma_1 = \gamma_2$, the above bound is reduced to the parameters error bound (10.39). The bounds (10.39) show the strictest upper bounds that the approach can possess by assuming a rich exploration of the PE signal. The approximator (10.30) under the update rule (10.35) gives

$$\hat{S}_k = \phi_k^\mathsf{T}\theta_k + \alpha\phi_k^\mathsf{T}\left(r_{k+1} + \frac{q_k q_k^\mathsf{T}}{1 + q_k^\mathsf{T}q_k}\theta_k\right). \tag{10.45}$$

The term in brackets is bounded, and if it satisfies the PE condition (10.38), then the bounds (10.39) hold. Therefore, we can conclude the convergence of our reinforcement learning to a small bounded zone in terms of the residual error v_k, so $\delta_k \to v_k$. ∎

The following theorem shows how the PE signal affects the parameters error.

Theorem 10.4 *If the learning input $q_k/(1 + q_k^T q_k)$ in (10.35) is PE, and the reinforcement learning based $\mathcal{H}_2$ neural control (10.17) and (10.37) is used, then the parameters converge into the following bounded residual set:*

$$\|\widetilde{\theta}_k\| \leq \frac{\sqrt{\beta_2 T}\left[(1+\gamma_1)\gamma_2 + \alpha\beta_2(\gamma_1 + \gamma_2)\right]}{\beta_1\gamma_1(1-\gamma_2)}\overline{v}. \tag{10.46}$$

Proof: The error dynamics (10.36) can be rewritten as the following linear time-variant system

$$\widetilde{\theta}_{k+1} = \alpha\frac{q_k}{q_k^T q_k + 1}u_k,$$
$$y_k = \frac{q_k^T}{q_k^T q + 1}\widetilde{\theta}_k, \tag{10.47}$$

where $u_k = -y_k - \frac{1}{q_k^T q_{k+1}}v_k$ is the output feedback controller. Also the system (10.47) can be expressed in the following form:

$$x_{k+1} = x_k + B_k u_k,$$
$$y_k = C_k^T x_k. \tag{10.48}$$

The state and output for any time instance T is given by

$$x_{k+T} = x_k + \sum_{i=k}^{k+T-1} B_i u_i,$$

$$y_{k+T} = C_{k+T}^T x_{k+T}.$$

Let C_{k+T} satisfy the PE condition (10.38) so that

$$\beta_1 I \leq K_1 = \sum_{j=k+1}^{k+T} C_j C_j^T \leq \beta_2 I.$$

Then the output of the system is

$$y_{k+T} = C_{k+T}^T x_k + C_{k+T}^T \sum_{i=k}^{k+T-1} B_i u_i.$$

For the PE condition,

$$\sum_{j=k+1}^{k+T} C_j C_j^T x_k = \sum_{j=k+1}^{k+T} C_j \left(y_j - C_j^T \sum_{i=k}^{k+T-1} B_i u_i \right)$$

$$K_1 x_k = \sum_{j=k+1}^{k+T} C_j \left(y_j - C_j^T \sum_{i=k}^{k+T-1} B_i u_i \right).$$

Then,

$$x_k = K_1^{-1} \left(\sum_{j=k+1}^{k+T} C_j \left(y_j - C_j \sum_{i=k}^{k+T-1} B_i u_i \right) \right)$$

$$\|x_k\| \le \left\| K_1^{-1} \sum_{j=k+1}^{k+T} C_j y_j \right\| + \| K_1^{-1} \sum_{j=k+1}^{k+T} C_j C_j^{\mathsf{T}} \sum_{i=k}^{k+T-1} B_i u_i \|$$

$$\le \beta_1^{-1} \left\| \sum_{j=k+1}^{k+T} C_j C_j^{\mathsf{T}} \right\| 1/2 \left\| \sum_{j=k+1}^{k+T} y_j^{\mathsf{T}} y_j \right\|^{1/2} + \beta_1^{-1} \left\| \sum_{j=k+1}^{k+T} C_j C_j^{\mathsf{T}} \right\| \left\| \sum_{i=k}^{k+T-1} B_i u_i \right\|$$

$$\le \frac{\sqrt{\beta_2 T}}{\beta_1} \|y_k\| + \frac{\beta_2}{\beta_1} \sum_{i=k}^{k+T-1} \|B_i\| \|u_i\|.$$

Let $B_k = \alpha \frac{q_k}{q_k^{\mathsf{T}} q_k + 1}$ and $C_k = \frac{q_k}{q_k^{\mathsf{T}} q_k + 1}$ and $x_k = \widetilde{\theta}_k$,

$$\|u_k\| \le \|y_k\| + \| \frac{q_k}{q_k^{\mathsf{T}} q_k + 1} v_k \| \le \|y_k\| + \|v_k\|.$$

So

$$\sum_{i=k}^{k+T-1} \|B_i\| \|u_i\| = \sum_{i=k}^{k+T-1} \| \alpha \frac{q_i}{q_i^{\mathsf{T}} q_i + 1} \| \|u_i\|$$

$$\le \alpha(\|y_k\| + \|v_k\|) \sum_{i=k}^{k+T-1} \| \frac{q_i}{q_i^{\mathsf{T}} q_i + 1} \|$$

$$\le \alpha(\|y_k\| + \|v_k\|) \sqrt{\sum_{i=k}^{k+T-1} \| \frac{q_i}{q_i^{\mathsf{T}} q_i + 1} \|} \sqrt{\sum_{i=k}^{k+T-1} 1}$$

$$\le \alpha(\|y_k\| + \|v_k\|) \sqrt{\beta_2 T}.$$

Finally

$$\|\widetilde{\theta}_k\| \le \frac{\sqrt{\beta_2 T}}{\beta_1} \left(\|y_k\| + \alpha \beta_2 (\|y_k\| + \|v_k\|) \right). \tag{10.49}$$

The system output y_k is a normalization of the approximator $F(\theta)$ in (10.30) such that it satisfies

$$\|y_k\| = \| \frac{q_k^{\mathsf{T}}}{q_k^{\mathsf{T}} q_k + 1} (\theta - \theta^*) \| \le \|F(\theta) - F(\theta^*)\|.$$

Since $q_k^{\mathsf{T}} q_k + 1 \ge 1$, we have

$$\|y_k\| \le \frac{\gamma_2(1 + \gamma_1)}{\gamma_1(1 - \gamma_2)} \bar{v},$$

and the residual error $\|v_k\| \le \bar{v}$. Substituting the output and residual error upper bounds into (10.49),

$$\widetilde{\theta}_k \le \frac{\sqrt{\beta_2 T} \left[(1 + \gamma_2)\gamma_2 + \alpha \beta_2(\gamma_2 + \gamma_2) \right]}{\beta_1 \gamma_2(1 - \gamma_2)} \bar{v}.$$

If the number of neurons at the hidden layer are increased, i.e, $p \to \infty$, then the residual error is decreased ($v_k \to 0$), and therefore the temporal difference error $\delta_k \to v_k$. Nevertheless, this can cause the overfitting problem at the training phase of the neural network approximator. ∎

Remark 10.3 *Similarly to adaptive controllers and identification algorithms, the PE signal must be designed carefully such that (10.38) is satisfied. A simple way to construct the PE signal is by adding the zero-mean exploration term Δu_k at the control input. Some authors use ε -greedy exploration as the PE signal, where the controller explores the input space until it finds the optimal or sub-optimal solution. The controller need to search the equilibrium between exploring and exploiting the knowledge in the reinforcement learning. We use the zero-mean exploration method by adding some signals with different frequency [27] to the desired reference.*

The $\mathcal{H}_2$ neural control using reinforcement learning is given in Algorithm 10.2, which is somewhat similar to the one presented in Algorithm 10.1.

Algorithm 10.2 RL $\mathcal{H}_2$ control.

1: **print** Matrices A, Q^i, Q^c, R^c, $W_{1,0}$, $W_{2,0}$, θ_0. Scalars β, η_0, α, and γ. Activation functions $\sigma(\cdot)$ and $\phi(\cdot)$. Desired reference $\varphi(x^d)$.
2: Compute the kernel matrix P^i using (10.12)
3: **for** each step $k = 0, 1, \ldots$ **do**
4:     Do steps 5-9 of Algorithm 10.1. Measure x_{k+1} and calculate e_{k+1}
5:     Apply the feedback-feedforward control U_1 and U_2 using (10.17) and (10.37)
6:     Update the neural states θ_0 using (10.35)
7: **end for**

10.3 $\mathcal{H}_2$ Neural Control in Continuous Time

Consider the following unknown continuous-time nonlinear system

$$\dot{x}_t = f(x_t, u_t), \tag{10.50}$$

where $x_t \in \mathbb{R}^n$ is the state vector, $u_t \in \mathbb{R}^m$ is the control input, and $f(x_t, u_t) : \mathbb{R}^n \times \mathbb{R}^m \to \mathbb{R}^n$ defines the system nonlinear dynamics.

We use the following differential neural network to model the nonlinear system (10.50):

$$\dot{\hat{x}}_t = A\hat{x}_t + W_t \sigma(V_t \hat{x}_t) + U_t, \tag{10.51}$$

where $\hat{x}_t \in \mathbb{R}^n$ is the state vector of the neural network, $A \in \mathbb{R}^{n \times n}$ is a Hurwitz matrix, $W_t \in \mathbb{R}^{n \times r}$ is the weight matrix of the output weights, $V_t \in \mathbb{R}^{k \times n}$ is the weight matrix of the hidden layer, and $U_t = [u_{1,t}, u_{2,t}, \cdots, u_{m,t}, 0, \cdots, 0]^{\mathsf{T}} \in \mathbb{R}^n$ is the control action.

The elements of $\sigma(\cdot) : \mathbb{R}^k \to \mathbb{R}^r$ increase monotonically and are chosen as sigmoid functions, $\sigma_i(x) = \frac{a_i}{1+e^{-b_i x_i}} - c_i$. The identification error is defined as

$$\tilde{x}_t = \hat{x}_t - x_t. \tag{10.52}$$

The sigmoid function $\sigma(\cdot)$ satisfies the following generalized Lipshitz condition,

$$\tilde{\sigma}_t^{\mathsf{T}} \Lambda_1 \tilde{\sigma}_t \leq \tilde{x}_t^{\mathsf{T}} \Lambda_\sigma \tilde{x}_t,$$
$$\tilde{\sigma}_t'^{\mathsf{T}} \Lambda_1 \tilde{\sigma}_t' \leq \left(\tilde{V}_t \hat{x}_t \right)^{\mathsf{T}} \Lambda_V \tilde{V}_t \hat{x}_t,$$

where $\tilde{\sigma}_t = \sigma(V^* \hat{x}_t) - \sigma(V^* x_t)$ is the neural function difference, $\Lambda_1, \Lambda_\sigma$, and Λ_V are known normalizing positive constant matrices, and V^* can be considered as the initial values of the hidden layer weights.

This neural model is a parallel recurrent neural network. Using Taylor series around the points $V_t \hat{x}_t$, gives

$$\sigma(V_t \hat{x}_t) = \sigma(V^* \hat{x}_t) + \underbrace{D_\sigma \tilde{V}_t \hat{x}_t + \varepsilon_V}_{\tilde{\sigma}_t'},$$

where $\tilde{V}_t = V_t - V^*$, $D_\sigma = \frac{d\sigma(V_t \hat{x}_t)}{dV_t \hat{x}_t} \in \mathbb{R}^{r \times k}$, and ε_V is a second order approximation error, which satisfies

$$\|\varepsilon_V\| \leq L_1 \|\tilde{V}_t \hat{x}_t\|, \quad \|\varepsilon_V\|_{\Lambda_1}^2 \leq L_1 \|\tilde{V}_t \hat{x}_t\|_{\Lambda_1}^2, \quad L_1 > 0.$$

According to the Stone-Weierstrass theorem [25], the nonlinear model (10.50) can be written as

$$\dot{x}_t = A x_t + W^* \sigma(V^* x_t) + U_t, \tag{10.53}$$

where W^* is the optimal weight matrix, which is bounded as follows:

$$W^* \Lambda_1^{-1} W^{*\mathsf{T}} \leq \overline{W},$$

and $\overline{W}$ is a known matrix.

For nonlinear system identification, we consider the following two cases:

1) We assume that the neural model (10.51), with fixed V^*, can exactly approximate the nonlinear system (10.50). The identification error dynamics between (10.53) and (10.51) is

$$\dot{\tilde{x}}_t = A\tilde{x} + \tilde{W}_t \sigma(V^* \hat{x}_t) + W^* \tilde{\sigma}_t, \tag{10.54}$$

where $\tilde{W}_t = W_t - W^*$. Hence, the nonlinear model (10.50) can be written as

$$\dot{x}_t = A x_t + W_t \sigma(V^* \hat{x}_t) + U_t + d_t, \tag{10.55}$$

where $d_t = -W^* \tilde{\sigma}_t - \tilde{W}_t \sigma(V^* \hat{x}_t)$.

2) Generally, the neural network cannot fully describe the nonlinear system dynamics (10.50). Let us fix some weight matrices W^* and V^* and a stable matrix A. Define the unmodeled dynamics between (10.50) and (10.51) as

$$-\eta_t = Ax_t + W^*\sigma(V^*\hat{x}_t) + U_t - f(x_t, u_t). \tag{10.56}$$

For any positive definite matrix Λ_1, there exists a positive constant $\bar{\eta}$ such that

$$\|\eta_t\|^2_{\Lambda_1} = \eta_t^{\mathsf{T}}\Lambda_1\eta_t \leq \bar{\eta} < \infty, \quad \Lambda_1 = \Lambda_1^{\mathsf{T}} > 0,$$

where $\bar{\eta}$ is a known upper bound.

The identification error dynamics considering the modeling error η_t becomes

$$\dot{\tilde{x}}_t = A\tilde{x} + \widetilde{W}_t\sigma(V_t\hat{x}_t) + W^*(\tilde{\sigma}_t + \tilde{\sigma}'_t) - \eta_t. \tag{10.57}$$

The dynamics of the nonlinear system (10.50) can be rewritten as

$$\dot{x}_t = Ax_t + W_t\sigma(V_t\hat{x}_t) + U_t + \xi_t, \tag{10.58}$$

where $\xi_t = \eta_t - W^*(\tilde{\sigma}_t + \tilde{\sigma}') - \widetilde{W}_t\sigma(V_t\hat{x}_t)$.

We want to find the estimated model (10.51) that minimizes the following discounted value function:

$$V_1 = \int_t^\infty (\tilde{x}_\tau^{\mathsf{T}} S\tilde{x}_\tau)e^{t-\tau}d\tau, \tag{10.59}$$

where $S = S^{\mathsf{T}} > 0$ is an $n \times n$ strictly positive definite weight matrix. Here we use a discount factor $e^{t-\tau} = 1$. This discounted version helps to guarantee convergence of the cost function.

Since A is a Hurwitz matrix, there exist positive definite matrices $R_i = R_i^{\mathsf{T}} = 2\overline{W} + \Lambda_1^{-1}$ and $Q_i \triangleq S + \Lambda_\sigma$ such that the Riccati equation

$$-\dot{P}_i = A^{\mathsf{T}}P_i + P_iA + Q_i + P_iR_iP_i - P_i \tag{10.60}$$

has positive solution $P_i = P_i^{\mathsf{T}} > 0$.

We can write the value function (10.59) in quadratic form:

$$V_1 = \tilde{x}_t^{\mathsf{T}} P_i\tilde{x}_t. \tag{10.61}$$

Remark 10.4 *The stable matrix A assures that the Riccati equation (10.60) has positive solution P_i. Matrix A can be freely chosen such that it is stable. A is a tunable parameter of the neural identifier. Although it will affect the neural modeling accuracy, this change can be compensated by the reinforcement learning.*

Remark 10.5 *The unmodeled dynamic η_t in (10.57) is structure error, and it depends on the numbers of the nodes and hidden layer. We will use the reinforcement learning to learn the tracking error such that the neural modeling error is compensated by the reinforcement learning. So we do not pay attention to the design* the neural networks (10.51). To simplify the modeling process, we can even use the single layer neural network (V_t of (10.51) is fixed) and let the reinforcement learning to deal with the big modeling error.

The main control goal is to force the system states to track a desired trajectory $x_{d_t} \in \mathbb{R}^n$, which is assumed to be smooth. This trajectory is regarded as a solution of the following known nonlinear reference model:

$$\dot{x}_{d_t} = \varphi(x_{d_t}) \tag{10.62}$$

with a fixed and known initial condition. Here $\varphi(\cdot)$ denotes the dynamics of the reference model. The initial condition of (10.62) is known as x_{d_0}.

We define the tracking error as

$$e_t = x_t - x_{d_t}. \tag{10.63}$$

The closed-loop error dynamics can be obtained using (10.63), (10.62), and either (10.55) or (10.58) as

$$\begin{aligned}
\dot{e}_t &= Ax_t + W_t \sigma(V^* \hat{x}_t) + U_t + d_t - \varphi(x_{d_t}) \\
\dot{e}_t &= Ax_t + W_t \sigma(V_t \hat{x}_t) + U_t + \xi_t - \varphi(x_{d_t}).
\end{aligned} \tag{10.64}$$

The neural $\mathcal{H}_2$ control is

$$\begin{aligned}
U_t &= U_1 + U_2, \\
U_1 &= \varphi(x_{d_t}) - Ax_d - W_t \sigma(V^* \hat{x}_t), \\
U_1 &= \varphi(x_{d_t}) - Ax_d - W_t \sigma(V_t \hat{x}_t),
\end{aligned} \tag{10.65}$$

where U_1 is a feedback linearization controller for (10.55) and (10.58), and U_2 is the $\mathcal{H}_2$ optimal control.

In case (1), the neural model (10.51) can exactly approximate the nonlinear system (10.50). The dynamics of the tracking error (10.63) under the feedback linearization controller U_1 becomes

$$\dot{e}_t = Ae_t + U_2 + d_t. \tag{10.66}$$

In case (2), the neural model (10.51) cannot approximate the nonlinear system (10.50) exactly, and the dynamics of the tracking error is reduced to

$$\dot{e}_t = Ae_t + U_2 + \xi_t. \tag{10.67}$$

In both cases, the controled system can be regarded as a linear system with disturbances.

We want to design the $\mathcal{H}_2$ optimal control U_2 that minimizes the following discounted cost function:

$$V_2 = \int_t^\infty (e_\tau^\top Q_c e_\tau + U_{2,\tau}^\top R_c U_{2,\tau}) e^{t-\tau} d\tau, \tag{10.68}$$

where $Q_c = Q_c^\mathsf{T} > 0$ and $R_c = R_c^\mathsf{T} > 0$ are weight matrices of the tracking error e_t and control U_2, respectively.

Because A is stable, there exists a positive definite kernel matrix $P_c = P_c^\mathsf{T} > 0$, which is the solution of the following Riccati equation:

$$-\dot{P}_c = A^\mathsf{T} P_c + P_c A + Q_c - P_c R_c^{-1} P_c - P_c. \tag{10.69}$$

(10.68) in quadratic form is

$$V_2 = e_t^\mathsf{T} P_c e_t. \tag{10.70}$$

For the linear part, the $\mathcal{H}_2$ optimal controller becomes LQR form

$$U_2 = -K e_t = -R_c^{-1} P_c e_t. \tag{10.71}$$

The following theorem gives the asymptotic stability of the neural $\mathcal{H}_2$ control when the neural model (10.51) can exactly approximate the nonlinear system (10.50).

Theorem 10.5 *If the nonlinear system (10.50) can be modeled exactly by the differential neural network (10.53) with fixed hidden weights V^*, W_t is updated as*

$$\dot{W}_t = -K_1 P_i \tilde{x}_t \sigma(V^* \hat{x}_t), \tag{10.72}$$

where $K_1 \in R^{n \times n}$ is positive definite, then the estimation error $\tilde{x}_t$ converges asymptotically to zero, and the tracking error e_t converges to a small bounded zone.

Proof: Define the following Lyapunov function

$$V = V_1 + V_2 + V_3, \tag{10.73}$$

where V_1 and V_2 are defined by (10.61) and (10.70), and V_3 is given by

$$V_3 = \mathrm{tr}\left[(\widetilde{W}_t)^\mathsf{T} K_1^{-1} \widetilde{W}_t \right]. \tag{10.74}$$

Taking the time derivative of V_1 along the identification error dynamics (10.54), the value function (10.59) gives (using Leibniz rule):

$$\dot{V}_1 = \tilde{x}_t^\mathsf{T} \left(\dot{P}_i + A^\mathsf{T} P_i + P_i A \pm S \pm P_i \right) \tilde{x}_t \\ + 2\tilde{x}_t^\mathsf{T} P_i \widetilde{W}_t \sigma^\mathsf{T} (V^* x_t) + 2\tilde{x}_t^\mathsf{T} P_i W^* \tilde{\sigma}_t.$$

The following inequality is used:

$$X^\mathsf{T} Y + \left(X^\mathsf{T} Y \right)^\mathsf{T} \leq X^\mathsf{T} \Lambda^{-1} X + Y^\mathsf{T} \Lambda Y, \tag{10.75}$$

where $X, Y \in \mathbb{R}^{n \times m}$ and $\Lambda = \Lambda^\mathsf{T} > 0$ is a positive definite matrix. Then $\dot{V}_1$ is rewritten as

$$\dot{V}_1 \leq \tilde{x}_t^\mathsf{T} \left(\dot{P}_i + 2A^\mathsf{T} P_i \pm P_i + P_i R_i P_i + \Lambda_\sigma \right) \tilde{x}_t + 2\tilde{x}_t^\mathsf{T} P_i \widetilde{W}_t \sigma^\mathsf{T} (V^* x_t),$$

where $R = \overline{W} + \Lambda_1^{-1}$. The time derivative of V_3 is

$$\dot{V}_3 = 2\mathrm{tr}\left[\widetilde{W}_t^T K_1^{-1} \dot{\widetilde{W}}_t\right].$$

The time derivative of V_2 along the trajectories of the tracking error (10.66), the value function (10.68) is (using Leibniz rule):

$$\dot{V}_2 = e_t^T \left(\dot{P}_c + A^T P_c + P_c A \pm Q_c \pm P_c\right) e_t$$
$$+ 2e_t^T P_c (U_2 + d_t) \pm U_2^T R_c U_2.$$

Substituting the control law (10.71), the differential Riccati equations (10.60), (10.69), and the training law (10.72) gives

$$\dot{V} = -\widetilde{x}_t^T \overline{Q}\widetilde{x}_t - e_t^T Q_x e_t + 2e_t^T P_c d_t^T, \tag{10.76}$$

where $\overline{Q} = Q_i - P_i - \Lambda_\sigma$ and $Q_x = Q_c + P_c R_c^{-1} P_c - P_c$. The first term of (10.76) is negative definite if $S > P_i$, and therefore, $Q_i > P_i + \Lambda_\sigma$, which is easy to fulfill by choosing a big enough S matrix. The second term of (10.76) is negative definite due to the design of the Riccati equation (10.69). From (10.76) we have that

$$\dot{V} \leq -\lambda_{\min}\left(\overline{Q}\right) \|\widetilde{x}_t\|^2. \tag{10.77}$$

So $\widetilde{W}_t$, σ, $\widetilde{\sigma} \in \mathcal{L}_\infty$ and

$$\int_t^\infty \lambda_{\min}\left(\overline{Q}\right) \|\widetilde{x}_\tau\|^2 d\tau \leq V - V_\infty < \infty.$$

Thus, $\widetilde{x}_t \in \mathcal{L}_2 \cap \mathcal{L}_\infty$ and $\dot{\widetilde{x}}_t \in \mathcal{L}_\infty$. By Barbalat's lemma [28], we can conclude that

$$\lim_{t \to \infty} \widetilde{x}_t = 0. \tag{10.78}$$

From (10.76), we have that

$$\dot{V} \leq -\left(\lambda_{\min}(Q_x)\|e_t\| - 2\lambda_{\max}(P_c)\|d_t\|\right)\|e_t\|. \tag{10.79}$$

Therefore $\dot{V} \leq 0$ as long as

$$\|e_t\| > \frac{2\lambda_{\max}(P_c)}{\lambda_{\min}(Q_x)}\|d_t\| \equiv \epsilon_0. \tag{10.80}$$

If we select the weight matrices Q_c and R_c such that $\lambda_{\min}(Q_x) > 2\lambda_{\max}(P_c)\|d_t\|$ is satisfied, it is ensured that the trajectories of the tracking error dynamics (10.66) converge to a compact set S_0 of radius ϵ_0, i.e., $\|e_t\| \leq \epsilon_0$ and hence, the trajectories of (10.66) are bounded. Furthermore, if $\widetilde{W}_t$ converges to zero as $t \to \infty$, then d_t also converges to zero, and therefore, e_t converges to zero. ∎

In the general case, the neural model (10.51) cannot approximate the nonlinear system (10.50) exactly. The following theorem gives the stability results of the H_2 neural control under modeling error.

Theorem 10.6 *The nonlinear system (10.50) is modeled by the differential neural network as in (10.51), if the weights are adjusted as*

$$\dot{W}_t = -sK_1 P_i \left[\sigma(V_t \hat{x}_t) - D_\sigma \tilde{V}_t \hat{x}_t \right] \hat{x}_t^\mathsf{T},$$

$$\dot{V}_t = -s \left[K_2 P_i W_t D_\sigma \tilde{x}_t \hat{x}_t^\mathsf{T} + \frac{L_2}{2} K_2 \Lambda_1 \tilde{V}_t \hat{x}_t \hat{x}_t^\mathsf{T} \right], \tag{10.81}$$

with

$$s = \begin{cases} 1 & \text{if } \|\tilde{x}_t\| > \sqrt{\dfrac{\bar{\eta}}{\lambda_{\min}(\bar{Q})}}, \\[3mm] 0 & \text{if } \|\tilde{x}_t\| \le \sqrt{\dfrac{\bar{\eta}}{\lambda_{\min}(\bar{Q})}}, \end{cases} \tag{10.82}$$

then the identification and tracking errors converge into bounded sets S_1 and S_2 of radius ϵ_1 and μ, respectively:

$$\lim_{t\to\infty} \|\tilde{x}_t\| \le \epsilon_1 = \sqrt{\frac{\bar{\eta}}{\lambda_{\min}(\bar{Q})}},$$

$$\lim_{t\to\infty} \|e_t\| \le \mu = \frac{2\lambda_{\max}(P_c)\|\xi_t\|}{\lambda_{\min}(Q_x)}.$$

Proof: The same Lyapunov function (10.73) is used, where V_3 is modified as follows:

$$V_3 = \text{tr}\left[(\tilde{W}_t)^\mathsf{T} K_1^{-1} \tilde{W}_t \right] + \text{tr}\left[(\tilde{V}_t)^\mathsf{T} K_2^{-1} \tilde{V}_t \right]. \tag{10.83}$$

The time derivative of V_1 along the identification error dynamics (10.57) is

$$\dot{V}_1 = \tilde{x}_t^\mathsf{T} \dot{P}_i \tilde{x}_t + \pm \tilde{x}_t^\mathsf{T} S \tilde{x}_t \pm \tilde{x}_t^\mathsf{T} P_i \tilde{x}_t$$
$$+ 2\tilde{x}_t^\mathsf{T} P_i (A\tilde{x}_t + \tilde{W}_t \sigma(V_t \hat{x}_t) + W^*(\tilde{\sigma}_t + \tilde{\sigma}_t') + \eta_t).$$

The time derivative of V_3 is

$$\dot{V}_3 = 2\text{tr}\left[\tilde{W}_t^\mathsf{T} K_1^{-1} \dot{\tilde{W}}_t \right] + 2\text{tr}\left[\tilde{V}_t^\mathsf{T} K_2^{-1} \dot{\tilde{V}}_t \right].$$

Using the matrix inequality (10.75) and substituting the updating law (10.81) in $\dot{V}_1 + \dot{V}_3$ gives the following cases:

- If $\|\tilde{x}_t\| > \sqrt{\bar{\eta}/\lambda_{\min}(\bar{Q})}$,

$$\dot{V}_1 \le \tilde{x}_t^\mathsf{T} \left(\dot{P}_i + 2P_i A \pm Q_i + P_i R_i P_i \pm P_i + \Lambda_\sigma \right) \tilde{x}_t + \bar{\eta}_t$$
$$= -\tilde{x}_t^\mathsf{T} \bar{Q} \tilde{x}_t + \bar{\eta} \le -\lambda_{\min}(Q)\|\tilde{x}_t\|^2 + \bar{\eta},$$

where $R_i = 2\bar{W} + \Lambda_1^{-1}$.

- If $\|\tilde{x}_t\| \le \sqrt{\bar{\eta}/\lambda_{\min}(\bar{Q})}$, W_t is constant. Therefore, V_3 is constant, and $\tilde{x}_t$ and W_t are bounded. So there exists a large enough matrix $\bar{Q}$ such that $\lambda_{\min}(\bar{Q}) \ge \bar{\eta}$, and

$\tilde{x}_t$ converges into a ball S_1 of radius ϵ_1 as $t \to \infty$, i.e.,

$$\lim_{t \to \infty} \|\tilde{x}_t\| \le \epsilon_1 = \sqrt{\frac{\bar{\eta}}{\lambda_{\min}(\bar{Q})}}.$$

Applying the control laws (10.65) and (10.71) in the time derivative of V_2 along the tracking error dynamics (10.67) gives

$$\dot{V}_2 = -e_t^\mathsf{T} Q_x e_t + 2e_t^\mathsf{T} P_c \xi_t$$
$$\le -\lambda_{\min}(Q_x)\|e_t\|^2 + 2\lambda_{\max}(P_c)\|e_t\|\|\xi_t\|$$
$$= -\|e_t\| \left[\lambda_{\min}(Q_x)\|e_t\| - 2\lambda_{\max}(P_c)\|\xi_t\| \right].$$

The above expression implies that there exists a large enough matrix Q_x such that $\lambda_{\min}(Q_x) > \lambda_{\max}(P_c)\|\xi_t\|$ and the tracking error converges into a bounded set S_2 of radius μ as $t \to \infty$, i.e.,

$$\lim_{t \to \infty} \|e_t\| \le \mu = \frac{2\lambda_{\max}(P_c)\|\xi_t\|}{\lambda_{\min}(Q_x)}. \qquad \blacksquare$$

Remark 10.6 *The dead-zone method in equations (10.81) and (10.82) is a way to deal with parameters drift in presence of modeling error [20]. If the neural identifier (10.51) does not have modeling error ($\eta = 0$), then by means of Theorem 1, $\tilde{x}$ converges to zero as $t \to \infty$.*

The $\mathcal{H}_2$ controller U_2 does not need any information of the real system (10.50). It assumes that all nonlinear dynamics are compensated by the neural controller U_1. The neural control U_1 is sensitive to the modeling error η_t. U_1 may have big tracking control error with even small modeling error η_t.

Since reinforcement learning does not need the dynamic model, we use reinforcement learning for the $\mathcal{H}_2$ control design. It can overcome the modeling error problem by using only measurements of the error state e_t and control U_2.

The continuous-time reinforcement learning uses the following discounted cost function:

$$Q(e_t, u_t) = \int_t^\infty (e_\tau^\mathsf{T} Q_c e_\tau + u_\tau^\mathsf{T} R_c u_\tau) e^{t-\tau} d\tau \qquad (10.84)$$

with any critic-learning methods [18], such as Q-learning or Sarsa. The action-value function (10.84) is equivalent to the Lyapunov value function V_2 in (10.83). The above critic methods use the Bellman optimality principle

$$u_t^* = \arg \min_u Q^*(e_t, u)$$

to obtain the $\mathcal{H}_2$ controller, but they need to explore all the possible state-action combinations. The learning time is large. The neural control (10.65) can accelerate the learning process by using the differential neural network (10.51).

We use the following neural approximator to approximate the value function V_2 in (10.68),

$$\hat{V}_2 = \theta_t^\mathsf{T} \phi(e_t) = \theta_t^\mathsf{T} \phi, \tag{10.85}$$

where $\theta_t \in \mathbb{R}^p$ is the weight vector, and $\phi(\cdot) : \mathbb{R}^n \to \mathbb{R}^p$ is the activation function vector with p neurons in the hidden layer.

The value function V_2 can be written as

$$V_2 = \theta^{*\mathsf{T}} \phi + \varepsilon(e_t), \tag{10.86}$$

where θ^* is an optimal weight value, and $\varepsilon(e_t)$ is the approximation error. The time derivative of the value function (10.68) is

$$\dot{V}_2 = -e_t^\mathsf{T} Q_c e_t - U_2^\mathsf{T} R_c U_2,$$

where we can write the derivative as $\dot{V}_2 = (\partial V_2 / \partial e_t) \dot{e}_t$. The partial derivative of the value function (10.86) is:

$$\frac{\partial V_2}{\partial e_t} = \nabla \phi^\mathsf{T} \theta^* + \nabla \varepsilon(e_t),$$

where $\nabla = \partial / \partial e_t$.

The Hamiltonian of the problem using the neural approximator (10.86) is

$$H(e_t, U_2, \theta^*) = \theta^{*\mathsf{T}}(\nabla \phi \dot{e}_t - \phi) + r_t = v_t$$
$$r_t = e_t^\mathsf{T} Q_c e_t + U_2^\mathsf{T} R_c U_2,$$

where $v_t = -\nabla^\mathsf{T} \varepsilon(e_t) \dot{e} + \varepsilon(e_t)$ is the residual error of the neural approximator, which is bounded, and r_t is defined as the utility function. The reinforcement learning method will minimize the residual error.

Consider the approximation of the Hamiltonian

$$\hat{H}(e_t, U_2, \theta_t) = \theta_t^\mathsf{T}(\nabla \phi \dot{e}_t - \phi) + r_t = \delta_t, \tag{10.87}$$

where δ_t is the temporal difference error (TD error) of the reinforcement learning. It can be written as

$$\delta_t = \tilde{\theta}_t^\mathsf{T}(\nabla \phi \dot{e}_t - \phi) + v_t,$$

where $\tilde{\theta}_t = \theta_t - \theta^*$.

The objective function of the reinforcement learning is to minimize the squared TD error

$$E = \frac{1}{2}\delta_t^2.$$

We use the normalized gradient descent algorithm,

$$\dot{\theta}_t = -\alpha \delta_t \frac{q_t^\mathsf{T}}{(q_t^\mathsf{T} q_t + 1)^2}, \tag{10.88}$$

where $\alpha \in (0, 1]$ is the learning rate and

$$q_t = \nabla \phi \dot{e}_t - \phi,$$

This normalized gradient descent assures that the weights are bounded [11].

The update rule (10.88) can be rewritten as

$$\dot{\tilde{\theta}}_t = -\alpha \frac{q_t q_t^T}{(q_t^T q_t + 1)^2} \tilde{\theta}_t - \alpha \frac{q_t}{(q_t^T q_t + 1)^2} v_t. \tag{10.89}$$

The $\mathcal{H}_2$ controller is obtained by solving the stationary condition

$$\partial \hat{H} / \partial U_2 = 0.$$

So

$$U_2 = -\frac{1}{2} R_c^{-1} \nabla \phi^T (e_t) \theta_t. \tag{10.90}$$

Substituting (10.90) into the Hamiltonian, the Hamilton-Jacobi-Bellman equation is

$$\delta_t = e_t^T Q_c e_t + \theta_t^T [\nabla \phi (A e_t + \xi_t) - \phi] - \frac{1}{4} \theta_t^T \nabla \phi R_c^{-1} \nabla \phi^T \theta_t.$$

Convergence of the reinforcement learning algorithm implies that $\hat{H}(\theta_t)$ converges to $H(\theta^*)$, when the TD error δ_t converges to v_t as $t \to \infty$ and hence, $\theta \to \theta^*$. This result defines the convergence property of the $\mathcal{H}_2$ control with RL.

We first define the following persistent exciting (PE) condition.

Definition 10.2 *The signal $q_t / (q_t^T q_t + 1)$ in (10.88) is persistently exciting (PE) over the time interval $[t, t + T]$, if there exists constants $\beta_1, \beta_2 > 0, T > 0$ such that for all time t the following holds:*

$$\beta_1 I \leq L_0 = \int_t^{t+T} \frac{q_\gamma q_\gamma^T}{(q_\gamma^T q_\gamma + 1)^2} d\gamma \leq \beta_2 I. \tag{10.91}$$

The following theorem shows the weights convergence and the convergence of TD error δ_t.

Theorem 10.7 *If the learning input $q_t / (q_t^T q_t + 1)$ in (10.88) is persistently exciting as in (10.91), then the neural approximation error $\tilde{\theta}_t = \theta_t - \theta^*$ converges into a bounded set*

$$\|\tilde{\theta}_t\| \leq \frac{\sqrt{\beta_2 T}}{\beta_1} (1 + 2\alpha \beta_2) \bar{v},$$

where $\bar{v}$ is an upper bound of the residual error v_t.

Proof: Consider the following Lyapunov function

$$V_4 = \frac{1}{2} \, \text{tr} \left[\tilde{\theta}_t^\mathsf{T} \alpha^{-1} \tilde{\theta}_t \right]. \tag{10.92}$$

The time derivative of V_4 is given by

$$\dot{V}_4 = \text{tr} \left[\dot{\tilde{\theta}}_t^\mathsf{T} \alpha^{-1} \dot{\tilde{\theta}}_t \right]$$

$$= -\text{tr} \left[\tilde{\theta}_t^\mathsf{T} \frac{q_t q_t^\mathsf{T}}{(q_t^\mathsf{T} q_t + 1)^2} \tilde{\theta}_t \right] - \text{tr} \left[\tilde{\theta}_t^\mathsf{T} \frac{q_t}{(q_t^\mathsf{T} q_t + 1)^2} v_t \right]$$

$$\leq - \left\| \frac{q_t^\mathsf{T}}{q_t^\mathsf{T} q_t + 1} \tilde{\theta}_t \right\|^2 + \left\| \frac{q_t^\mathsf{T}}{q_t^\mathsf{T} q_t + 1} \tilde{\theta}_t \right\| \left\| \frac{v_t}{q_t^\mathsf{T} q_t + 1} \right\|$$

$$\leq - \left\| \frac{q_t^\mathsf{T}}{q_t^\mathsf{T} q_t + 1} \tilde{\theta}_t \right\| \left[\left\| \frac{q_t^\mathsf{T}}{q_t^\mathsf{T} q_t + 1} \tilde{\theta}_t \right\| - \left\| \frac{v_t}{q_t^\mathsf{T} q_t + 1} \right\| \right].$$

The update (10.89) can be rewritten as a linear time-variant (LTV) system

$$\dot{\tilde{\theta}}_t = -\alpha \frac{q_t q_t^\mathsf{T}}{(q_t^\mathsf{T} q_t + 1)^2} \tilde{\theta}_t - \alpha \frac{q_t}{(q_t^\mathsf{T} q_t + 1)^2} v_t,$$

$$y_t = \frac{q_t^\mathsf{T}}{q_t^\mathsf{T} q_t + 1} \tilde{\theta}_t.$$

For any time interval $[t, t+T]$, the time derivative of the parameters error satisfies

$$\int_t^{t+T} \dot{\tilde{\theta}}_\tau d\tau = \tilde{\theta}_{t+T} - \tilde{\theta}_t.$$

Then the solution of the above system in a time interval $[t, t+T]$ is

$$\tilde{\theta}_{t+T} = \tilde{\theta}_t - \alpha \int_t^{t+T} \left(\frac{q_\tau q_\tau^\mathsf{T}}{(q_\tau^\mathsf{T} q_\tau + 1)^2} \tilde{\theta}_\tau + \frac{q_\tau v_\tau}{(q_\tau^\mathsf{T} q_\tau + 1)^2} \right) d\tau,$$

$$y_{t+T} = \frac{q_{t+T}^\mathsf{T}}{q_{t+T}^\mathsf{T} q_{t+T} + 1} \tilde{\theta}_{t+T}.$$

We define $C_t = \frac{q_t}{(q_t^\mathsf{T} q_t + 1)^2}$. Then

$$y_{t+T} = C_{t+T}^\mathsf{T} \tilde{\theta}_t - \alpha \int_t^{t+T} C_{t+T}^\mathsf{T} \left(C_\tau C_\tau^\mathsf{T} \tilde{\theta}_\tau + C_\tau v_\tau \right) d\tau.$$

Using the PE condition (10.91),

$$\int_t^{t+T} C_\gamma \left(y_\gamma + \alpha \int_t^\gamma C_\gamma^\mathsf{T} \left(C_\tau C_\tau^\mathsf{T} \tilde{\theta}_\tau + C_\tau v_\tau \right) d\tau \right) d\gamma$$
$$= \int_t^{t+T} C_\gamma C_\gamma^\mathsf{T} \tilde{\theta}_t d\gamma = L_0 \tilde{\theta}_t.$$

Therefore the parameters error solution is

$$\tilde{\theta}_t = L_0^{-1} \int_t^{t+T} C_\gamma \left[y_\gamma + \alpha \int_t^\gamma C_\gamma^\mathsf{T} \left(C_\tau (C_\tau^\mathsf{T} \tilde{\theta}_\tau + v_\tau) \right) d\tau \right] d\gamma. \tag{10.93}$$

Taking the norm in both sides of (10.93) gives

$$\|\tilde{\theta}_t\| \leq \left\| L_0^{-1} \int_t^{t+T} C_\gamma y_\gamma \, d\gamma \right\| + \left\| \alpha L_0^{-1} \int_t^{t+T} C_\gamma C_\gamma^\mathsf{T} \int_t^\gamma C_\tau (C_\tau^\mathsf{T} \tilde{\theta}_\tau + v_\tau) d\tau d\gamma \right\|$$

$$\leq \frac{1}{\beta_1} \left(\int_t^{t+T} C_\gamma C_\gamma^\mathsf{T} \, d\gamma \right)^{1/2} \left(\int_t^{t+T} y_\gamma y_\gamma^\mathsf{T} \, d\gamma \right)^{1/2}$$

$$+ \frac{\alpha}{\beta_1} \int_t^{t+T} C_\gamma C_\gamma^\mathsf{T} \, d\gamma \int_t^{t+T} C_\tau (C_\tau^\mathsf{T} \tilde{\theta}_\tau + v_\tau) d\tau.$$

The output of the LTV system is bounded by

$$\left\| \frac{q_t^\mathsf{T}}{q_t^\mathsf{T} q_t + 1} \tilde{\theta}_t \right\| > \bar{v} > \left\| \frac{v_t}{q_t^\mathsf{T} q_t + 1} \right\|, \qquad (10.94)$$

so $\dot{V}_4 \leq 0$. From (10.94) we obtain an upper bound for the parameters error in terms of the upper bound of the residual error as

$$\|\tilde{\theta}_t\| \leq \frac{\sqrt{\beta_2 T}}{\beta_1} \bar{v} + \frac{2\alpha\beta_2}{\beta_1} \sqrt{\beta_2 T \bar{v}} \leq \frac{\sqrt{\beta_2 T}}{\beta_1} (1 + 2\alpha\beta_2) \bar{v}.$$

This means that

$$\theta_t \to \theta^*, \quad \delta_t \to v \text{ as } t \to \infty. \qquad \blacksquare$$

Remark 10.7 *Using the TD error (10.87), we can obtain the following bound for the tracking error*

$$\|e_t\| \leq \frac{-b + \sqrt{b^2 + 4ac}}{2a}, \qquad (10.95)$$

where

$$a = \lambda_{\min}(Q_c), \quad b = \lambda_{\max}(A) \|\theta_t^\mathsf{T} \nabla \phi\|,$$
$$c = \|\delta_t\| + \tfrac{1}{4} \lambda_{\max}(R_c^{-1}) \|\theta_t^\mathsf{T} \nabla \phi\|^2 - \|\theta_t^\mathsf{T} \nabla \phi\| \|\xi_t\| - \|\theta_t^\mathsf{T} \phi\|.$$

The simplest NN approximator for (10.85) has the following quadratic form:

$$V_2 = \theta_t^\mathsf{T} \phi = e_t^\mathsf{T} P_{NN} e_t$$

for some kernel matrix $P_{NN} > 0$. The bound (10.95) is

$$\|e_t\| \leq \frac{2\lambda_{\max}(P_{NN}) \|\xi_t\|}{\lambda_{\min}(Q_c + P_{NN} R_c^{-1} P_{NN})}.$$

Hence, the upper bound of the tracking error depends on the accuracy of the NN approximator. If the number of neurons at the hidden layer is increased, i.e., $p \to \infty$, then the residual error is decreased, $v_t \to 0$, and therefore $\delta_t \to 0$, which implies the attenuation of the modeling error ξ_t. Nevertheless, this can cause the overfitting problem.

10.4 Examples

Example 10.1 *2-DOF Pan and Tilt Robot*

We use a 2-DOF pan and tilt robot (see Appendix A) to test the discrete-time $\mathcal{H}_2$ neural control. After a Euler discretization of the robot dynamics, the discrete time plant model can be written as

$$x^1_{1,k+1} = x^1_{1,k} + x^2_{1,k}T,$$

$$x^1_{2,k+1} = x^1_{2,k} + x^2_{2,k}T,$$

$$x^2_{1,k+1} = x^2_{1,k} + \left(\frac{4(\tau_{1,k} + \frac{1}{4}m_2 l_1^2 \dot{q}_1 \dot{q}_2 \sin(2q_2))}{m_2 l_2^2 \cos^2(q_2)} \right) T,$$

$$x^2_{2,k+1} = x^2_{2,k} + \left(\frac{8\tau_{2,k} - 2m_2 l_2(2g + l_1 \dot{q}_2^2)\cos(q_2) - m_2 l_1^2 \dot{q}_1^2 \sin(2q_2)}{2(m_2(l_1^2 + l_2^2) + 2m_2 l_1 l_2 \sin(2q_2))} \right) T,$$

$$(10.96)$$

where T is the sampling time, $\tau_{1,k}$ and $\tau_{2,k}$ are the applied torques, $x^1_i = q_i$, and $x^2_i = \dot{q}_i$, $i = 1, 2$, are the velocities.

Remark 10.8 *The system (10.96) is used only for simulation and its structure is used to desgin the neural network identifier. The parameters of the system (10.96) are assumed to be unknown for the control synthesis.*

We use two independent RNN (10.2) to estimate the position and velocity of each DOF. The first RNN is given by

$$\hat{x}^1_{k+1} = A\hat{x}^1_k + W^1_{1,k}\sigma(W^1_{2,k}\hat{x}^1_k) + \tau_k.$$

The second RNN is

$$\hat{x}^2_{k+1} = A\hat{x}^2_k + W^2_{1,k}\sigma(W^2_{2,k}\hat{x}^2_k) + \tau_k$$

where $\hat{x}^1_k = [\hat{x}^1_1, \hat{x}^1_2]^T$ and $\hat{x}^2_k = [\hat{x}^2_1, \hat{x}^2_2]^T$. We select the following values for the neural network identifiers:

$$A = -0.5I_{2\times2}, \quad Q^i = \begin{bmatrix} 5 & 2 \\ 2 & 5 \end{bmatrix}, \quad \beta = 1, \quad \sigma(x) = \tanh(x), \quad P^i_0 = 0.1I_{2\times2},$$

and 8 hidden nodes are used at the hidden layer and one node at the output layer. The initial conditions of the neural weights $W^1_{1,0} = W^2_{1,0} = W^{1T}_{2,0} = W^{2T}_{2,0} \in \mathbb{R}^{2\times8}$ are random numbers in the interval $[0, 1]$. The sampling time is $T = 0.01$ seconds. The desired references is

$$x^d_k = \begin{bmatrix} \sin\left(\frac{\pi}{3}k\right) \\ \sin\left(\frac{\pi}{4}k\right) \end{bmatrix}.$$

The weights are updated by (10.26). The neural control (10.17) is given by

$$\tau_k = x_{k+1}^d - Ax_k^d - W_{1,k}^1 \sigma(W_{2,k}^1 \hat{x}_k^1) + U_2,$$

where U_2 is slightly modified as

$$U_2 = -(R^c + P^c)^{-1} P^c A(e_k^1 + e_k^2),\qquad(10.97)$$

where $e_k^1 = x_k^1 - x_k^d$ and $e_k^2 = x_k^2 - x_{k+1}^d$. This modification is because only the first RNN is used for the model compensation. The second term of (10.97) dampens the closed-loop system. Another helpful modification is the following:

$$U_2 = -(R^c + P^c)^{-1} P^c A e_k^1 - (x_k^2 - \hat{x}_k^2).\qquad(10.98)$$

Here the second term is used to compensate unmodeled dynamics by using the identification error of the second RNN. The weight matrices of the optimal control are

$$Q^c = \begin{bmatrix} 5 & 0 \\ 0 & 5 \end{bmatrix}, \quad R^c = \begin{bmatrix} 0.1 & 0 \\ 0 & 0.1 \end{bmatrix}, \quad P^c = 0.1 I_{2\times2}.$$

For the neural RL solution, we use four neurons with quadratic activation function, i.e., $\phi_k = [(e_{1,k}^1)^2, (e_{2,k}^1)^2, (e_{1,k}^2)^2, (e_{2,k}^2)^2]^T$ with a learning rate of $\alpha = 0.1$ and a discount factor of $\gamma_1 = 0.9$. The NN weights are random numbers in the interval $[0, 1]$. The weight parameters of the neural RL are: $Q^c = 5 I_{4\times4}$ and $R^c = 0.1 I_{2\times2}$. Here we compare the performance of the LQR solutions, LQR1 (10.97) and LQR2 (10.98), with the neural RL solution (10.37). Recall that $t = kT$.

The tracking results are given in Figure 10.1. The results show good tracking performance of both methods using the RNN identifier and the LQR or RL controllers. The main difference between the LQR and the RL solutions is their precisions. The LQR1 control is sensible to the modeling error and hinders an accurate reference tracking. The LQR2 control uses the modeling error of the second RNN as compensation of the unmodeled dynamics. This compensation improves the tracking performance, but the output signal may have oscillations. To overcome this problem we can add more damping to the closed-loop system using a control gain; nevertheless, we can excite the modeling error and lose precision. The neural RL controller takes into account the modeling error, so the output control policy is improved.

We used the mean squared error (MSE) to see the precision of each controller as follows:

$$\bar{x}_1^1 = \frac{1}{n}\sum_{i=1}^{n}\left(e_{1,i}^1\right)^2, \quad \bar{x}_2^1 = \frac{1}{n}\sum_{i=1}^{n}\left(e_{2,i}^1\right)^2.$$

The bar plot of the MSE of the last 5 seconds is given in Figure 10.2. Notice that the neural RL MSE is much smaller than the LQR error. Another advantage of the

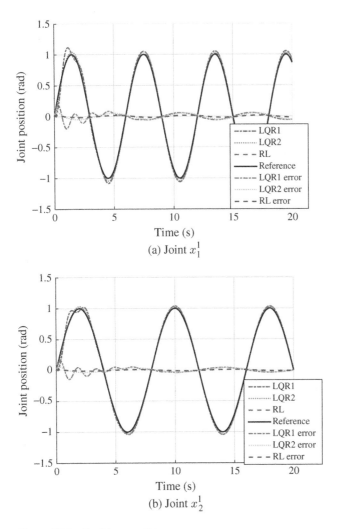

Figure 10.1 Tracking results.

neural RL approach is that we enhanced the robustness of the controller in presence of modeling error and convergence to an optimal or near-optimal control policy. On the other hand, the LQR solution is simple to design but it can only guarantee a local optimal performance, which is affected by disturbances or modeling error.

The Lyapunov value functions are quadratic in terms of the identification error $\tilde{x}_k^1$ and the tracking error e_k with respect to the kernel matrix and activation functions. The learning curves of the value functions V_k and S_k using RNN and RL are

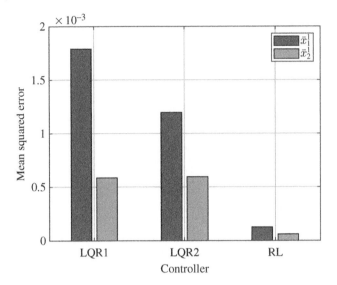

Figure 10.2 Mean squared error.

given in Figure 10.3(a) and Figure 10.3(b), respectively. The use of a discount factor is essential because if the reference trajectory does not go to zero, then the value function is infinite, i.e, if $\gamma_1 = 1$, then $\sum_{i=k}^{\infty}(e_i^T Q^c e_i + U_2^T R^c U_2) = \infty$.

The kernel matrices converge to the following values

$$P^i \leftarrow \begin{bmatrix} 10 & 4 \\ 4 & 10 \end{bmatrix}, \quad P^c \leftarrow \begin{bmatrix} 5.025 & 0 \\ 0 & 5.025 \end{bmatrix}.$$

In order to see the convergence of the elements of each kernel matrix we use the difference between each matrix in different time steps, i.e, $P_{k+1}(i,j) = P_k(i,j)$, where (i,j) stands for each element of the matrix. The convergence plot is shown in Figure 10.4.

It's worth noting that big values of matrices Q^i and Q^c get large kernel values. Small values of R^c do not noticeably affect the kernel solution. The eigenvalues of matrix A must be within the interval $(-1/\beta, 0)$ in order to guarantee convergence; otherwise, the kernel solution would diverge.

Example 10.2   *2-DOF Planar Robot*

The continuous-time $\mathcal{H}_2$ neural control is tested using a 2-DOF planar robot (see Appendix A). We use two independent RNN (10.51) to estimate the position and velocity of each DOF. The first RNN is given by

$$\dot{\hat{q}} = A\hat{q} + W_i \sigma(V_i\hat{q}) + \tau.$$

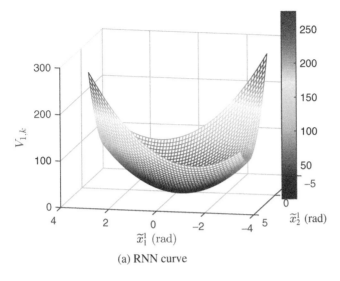

(a) RNN curve

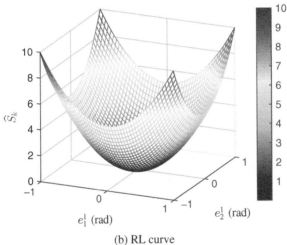

(b) RL curve

Figure 10.3 Learning curves.

The second RNN is

$$\ddot{\hat{q}} = A\dot{\hat{q}} + W_t'\sigma(V_t'\dot{\hat{q}}) + \tau,$$

where $q = [q_1, q_2]^\mathsf{T}$ and $\dot{q} = [\dot{q}_1, \dot{q}_2]^\mathsf{T}$. We select the following values for the neural network identifiers:

$$A = -12I_{2\times2}, \quad R_i = \begin{bmatrix} 8 & 2 \\ 2 & 8 \end{bmatrix}, \quad Q_i = \begin{bmatrix} 2 & 1 \\ 1 & 2 \end{bmatrix}, \quad \Lambda_1 = S = I_{2\times2}, \quad L_1 = 1$$

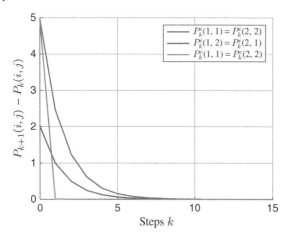

Figure 10.4 Convergence of kernel matrices P_k^i and P_k^c.

$$\sigma(x) = \frac{2}{1+e^{-2x}} - \frac{1}{2}, \quad W_0 = \begin{bmatrix} 2 & 0.5 & 1 \\ 0.1 & 0.5 & 0.75 \end{bmatrix}, \quad P_i^0 = \begin{bmatrix} 0.1 & 0.05 \\ 0.05 & 0.1 \end{bmatrix},$$

where I is the identity matrix and $W_0 = W_0' = V_0^{\mathsf{T}} = V_0'^{\mathsf{T}}$. The desired joint position and velocity references are

$$q_d = \begin{bmatrix} \cos\left(\frac{\pi}{4}t\right) \\ \sin\left(\frac{\pi}{6}t\right) \end{bmatrix}, \quad \dot{q}_d = \begin{bmatrix} -\frac{\pi}{4}\sin\left(\frac{\pi}{4}t\right) \\ \frac{\pi}{6}\cos\left(\frac{\pi}{6}t\right) \end{bmatrix}.$$

The weights are updated by (10.81); here $K_i = 15I$ ($i = 1, 2$). The neuro control (10.65) is given by

$$\tau = \dot{q}_d - Aq_d - W_t\sigma(V_t\hat{q}) + U_2,$$

where U_2 is slightly modified as follows:

$$U_2 = -R_c^{-1}P_c[(q - q_d) + (\dot{q} - \dot{q}_d)]. \tag{10.99}$$

This modification is due to the fact that only the first RNN is used for the model compensation. The velocity error is added to inject damping to the closed-loop system. Another way to inject damping to the closed-loop system is by means of the following control:

$$U_2 = -R_c^{-1}P_c(q - q_d) - (\dot{q} - \hat{\dot{q}}). \tag{10.100}$$

Here the second term is used to compensate unmodeled dynamics by using the velocity estimates of the second RNN. The weight matrices of the optimal control are

$$Q_c = \begin{bmatrix} 1 & 0 \\ 0 & 1 \end{bmatrix}, \quad R_c = \begin{bmatrix} 0.001 & 0 \\ 0 & 0.001 \end{bmatrix}, \quad P_c = 0.02I_{2\times2}.$$

Table 10.1 Parameters of the neural RL

Activation functions	State/Control weights	NN weights
$\phi(e) = \left[e_1^2,\ e_2^2,\ e_3^2,\ e_4^2\right]^{\mathsf{T}}$	$Q_c = I_{4\times4}\ R_c = 0.001I_{2\times2}$	$\theta_0 = \left[0.1,\ 0.5,\ 0.2,\ 0.4\right]^{\mathsf{T}}$

Figure 10.5 Tracking results.

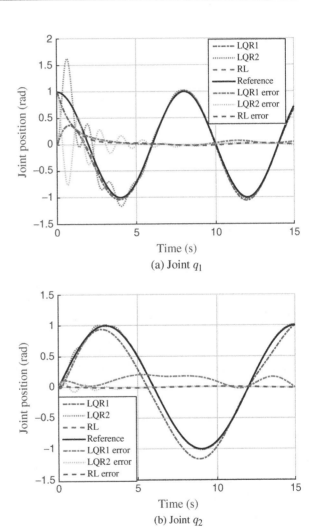

(a) Joint q_1

(b) Joint q_2

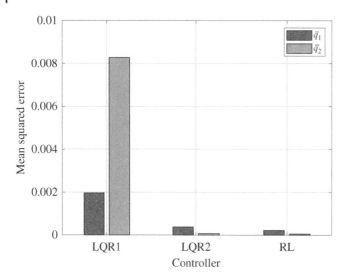

Figure 10.6 Mean squared error.

We compare the RNN solutions using the LQR control "LQR1" (10.99) and the LQR control "LQR2" (10.100) with the neural RL (10.89) denoted by "RL." The parameters of the neural RL are given in Table 10.1, where and $e_i, i = \overline{1,4}$, are the position and velocity errors of each DOF, respectively. The learning rate is $\alpha = 0.3$.

The tracking results of the neural reinforcement learning (RL) and LQR controls are given in Figure 10.5(a) and Figure 10.5(b). We can see that both methods attempt to follow the desired reference. The control LQR1 has good performance with small tracking error at the joint q_1; on the other hand, the control LQR1 shows a big tracking error at joint q_2 since the unmodeled dynamics affect the tracking performance. The control LQR2 shows oscillations due to a low damping effect. The modeling error term of the joint velocities helps to compensate the unmodeled dynamics and slightly inject damping. We can inject more damping at the closed-loop system by using a control gain; however, we can excite the modeling error and hinder an accurate reference tracking. The neural RL has good tracking performance since it takes into account both the joint positions and velocities of the robot and the modeling error.

Another advantage of the neural control is that the matrix A gives a feasible direction of the approximation and the neural estimator has fast convergence speed. A bad design of the functions $\sigma(\cdot)$ and $\phi(\cdot)$ may lead to large identification error, and therefore the tracking error increases. We use the mean squared error (MSE) as the performance metric of each controller. The MSE of the position

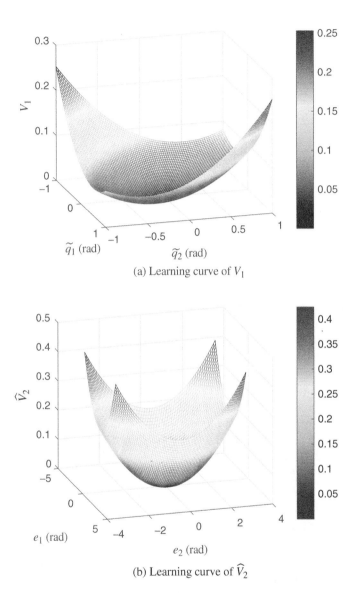

(a) Learning curve of V_1

(b) Learning curve of $\widehat{V}_2$

Figure 10.7 Learning curves.

tracking errors are given by

$$\overline{Q}_1 = \frac{1}{n}\sum_{i=1}^{n}(e_1(i)), \qquad \overline{Q}_2 = \frac{1}{n}\sum_{i=1}^{n}(e_2(i)).$$

The bar plot of the MSE error of the last 5 seconds is shown in Figure 10.6. The results show the advantage of the neural RL in comparison the LQR solutions.

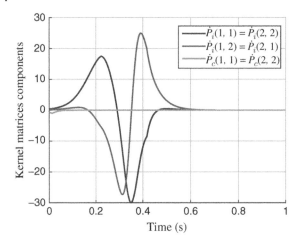

Figure 10.8 Convergence of kernel matrices $\dot{P}_i$ and $\dot{P}_c$

Control LQR1 shows a large MSE in comparison to LQR2 and RL. The main reason for this difference is the unmodeled dynamics. LQR2 has good MSE results, but this controller can present oscillations or excite the modeling error. The neural RL has good MSE results by taking into account measures of the position and velocity states and the modeling error.

The Lyapunov value functions are quadratic in terms of the identification error $\tilde{q}_t$ and the tracking error e_t with respect to the kernel matrix and neural activation functions. The learning curves of the Lyapunov value functions V_1 and V_2 using the Riccati equation (10.60) and RL (10.90) are shown in Figure 10.7(a) and Figure 10.7(b), respectively.

The kernel matrices converge (see Figure 10.8) to the following values:

$$
P_i \leftarrow \begin{bmatrix} 0.0834 & 0.043 \\ 0.043 & 0.0834 \end{bmatrix}, \quad P_c \leftarrow \begin{bmatrix} 0.0215 & 0 \\ 0 & 0.0215 \end{bmatrix}
$$

The selection of the matrix R_c helps to obtain a stable solution of the Riccati equation (10.69). Big values of the control weight R_i cause divergence of the kernel matrix P_i because the update (10.60) has a quadratic term that grows faster than the linear terms. Both LQR and RL solutions present good behavior; however, LQR is sensitive to the modeling error. The RL solution avoids this problem by taking the modeling error into account at its NN weights update. Thus, the LQR control is reliable when the modeling error does not affect the tracking precision. If the tracking precision depends on modeling accuracy, it is better to use the RL controller.

10.5 Conclusion

In this chapter, the $\mathcal{H}_2$ neural control for robots and nonlinear systems is proposed in both discrete and continuous time. The critic-learning method is based on differential neural networks, which serve for system identification and tracking control. The tracking control is solved using a feedforward and feedback controllers. The feedforward controllers realize the input-state linearization using the neural network identifier. The feedback controller is designed by two control techniques: a standard LQR control and a neural-based reinforcement learning. Stability and convergence are presented using Lyapunov stability theory and the contraction property. Simulations results show that the RL approach presents optimal and robust performances without knowledge of the system dynamics and large learning time.

References

1 F. Lewis and D. Vrabie, *"Optimal control,"* Wiley: New York, 2012.

2 F. Lewis, D. Vrable, and K. Vamvoudakis, "Reinforcement learning and feedback control using natural decision methods to desgin optimal adaptive controllers," *IEEE Control Systems Magazine*, 2012.

3 A. Al-Tamimi, F. Lewis, and M. Abu-Khalaf, "Model-free Q-learning designs for linear discrete-time zero-sum games with application to H_∞ control," *Automatica*, vol. 43, no. 3, pp. 473–481, 2007.

4 H. Zhang, D. Liu, Y. Luo, and D. Wang, *"Adaptive dynamic programming for control,"* Springer-Verlag: London, 2013.

5 R. Sutton and B. A, *Reinforcement Learning: An Introduction.* Cambridge, MA: MIT Press, 1998.

6 H. Modares and F. Lewis, "Linear quadratic tracking control of partially-unknown continuous-time systems using reinforcement learning," *IEEE Trans. Autom. Control*, vol. 59, no. 11, pp. 3051–3056, 2014.

7 I. Grondman, M. Vaandrager, L. Buşoniu, R. Babûska, and E. Schuitema, "Efficient model learning methods for actor-critic control," *IEEE Transactions on Systems, man, and cybernetics. Part B: Cybernetics*, vol. 42, no. 3, 2012a.

8 C. Wang, Y. Li, S. Sam Ge, and T. Heng Lee, "Optimal critic learning for robot control in time-varying environments," *IEEE Transactions on Neural Networks and Learning Systems*, vol. 26, no. 10, 2015.

9 A. Perrusquía, W. Yu, and A. Soria, "Large space dimension reinforcement learning for robot position/force discrete control," *2019 6th International Conference on Control, Decision and Information Technologies (CoDIT 2019)*, Paris, France, 2019.

10 K. Vamvoudakis, D. Vrabie, and F. Lewis, "Online learning algorithm for zero-sum games with integral reinforcement learning," *Journal of Artificial Intelligence and Soft Computing Research*, vol. 11, no. 4, pp. 315–332, 2011.

11 K. Vamvoudakis and F. Lewis, "Online actor-critic algorithm to solve the continuous-time infinite horizon optimal control problem," *Automatica*, vol. 46, pp. 878–888, 2010.

12 D. Vrabie, K. Vamvoudakis, and F. Lewis, "Optimal adaptive control and differential games by reinforcement learning principles," *Control, Robotics and Sensors, IET: Edison, New Jersey*, 2013.

13 I. Grondman, M. Vaandrager, L. Buşoniu, R. Babûska, and E. Schuitema, "Actor-critic control with reference model learning," *Proc. of the 18th World Congress The International Federation of Automatic Control*, pp. 14723–14728, 2011.

14 R. Kamalapurkar, P. Walters, and W. Dixon, "Model-based reinforcement learning for approximate optimal regulation," *Automatica*, vol. 64, pp. 94–104, 2016.

15 I. Grondman, L. Buşoniu, and R. Babûska, "Model learning actor-critic algorithms: performance evaluation in a motion control task," *51st IEEE Conference on Decision and Control (CDC)*, pp. 5272–5277, 2012b.

16 L. Dong, X. Zhong, C. Sun, and H. He, "Adaptive event-triggered control based on heuristic dynamic programming for nonlinear discrete-time systems," *IEEE Transactions on Neural Networks and Learning Systems*, vol. 28, no. 7, 2016.

17 A. Sahoo, H. Xu, and S. Jagannathan, "Near optimal event triggered control of nonlinear discrete-time systems using neurodynamic programming," *IEEE Transactions on Neural Networks and Learning Systems*, vol. 27, no. 9, pp. 1801–1815, 2016.

18 B. Kiumarsi, K. G. Vamvoudakis, H. Modares, and F. L. Lewis, "Optimal and autonomous control using reinforcement learning: A survey," *IEEE Transactions on Neural Networks and Learning Systems*, vol. 29, no. 6, 2018.

19 G. Leon and I. Lee, "Neural network indirect adaptive control with fast learning algorithm," *Neurocomputing*, vol. 12, no. 2-4, pp. 185–199, 1996.

20 W. Yu and A. Poznyak, "Indirect adaptive control via parallel dynamic neural networks," *IEEE Proceedings- Control Theory and Applications*, vol. 146, no. 1, pp. 25–30, 1999.

21 W. Yu and X. Li, "Discrete-time nonlinear system identification using recurrent neural networks," *Proceedings of the 42nd IEEE Conference on Decision and Control*, 2003, Hawai.

22 W. Yu, "Nonlinear system identification using discrete-time recurrent neural networks with stable learning algorithms," *Information Sciences*, vol. 158, pp. 131–147, 2004.

23 W. Yu and X. Li, "Recurrent fuzzy neural networks for nonlinear system identification," *22nd IEEE International Symposium on Intelligent Control Part of IEEE Multi-conference on Systems and Control*, 2007, Singapore.

24 W. Yu, "Multiple recurrent neural networks for stable adaptive control," *Neurocomputing*, vol. 70, pp. 430–444, 2006.

25 G. Cybenko, "Approximation by superposition of sigmoidal activation function," *Math, Control, Sig. Syst.*, vol. 2, pp. 303–314, 1989.

26 L. Buşoniu, R. Babûska, B. De Schutter, and D. Ernst, *Reinforcement learning and dynamic programming using function approximators*. CRC Press, Automation and Control Engineering Series, 2010.

27 H. Modares, F. Lewis, and Z. Jiang, "H_∞ tracking control of completely unknown continuous-time systems via off-policy reinforcement learning," *IEEE Transactions on Neural Networks and Learning Systems*, vol. 26, no. 10, pp. 2550–2562, 2015.

28 H. K. Khalil, *Nonlinear Systems*. Upper Saddle River, New Jersey: Prentice Hall, 3rd ed., 2002.

11

Conclusions

Robot interaction control has been maturing in recent years, including more difficult and sophisticated control problems such as in the medical robotics field. Novel techniques are developed for complex control task. This book analyzes robot-interaction control tasks using nonlinear model-based and model-free controllers. We discuss the robot-interaction control using reinforcement learning, which uses only stored data, or real-time data along the system trajectories.

The first part is on how robots work in an environment. Chapter 2 gives the concepts of impedance and admittance for environment modeling. Chapter 3 applies impedance and admittance control. Chapter 4 shows how to design model-free admittance control in joint space and task space.

The impedance and admittance models can be parameterized and estimated from the environment parameters identification. The impedance/admittance controller uses the feedback linearization to compensate the robot dynamics, and reduce the closed-loop system into a desired impedance model. The dilemma of accuracy and robustness arises with robot modeling. The model-free controllers overcome the accuracy and robustness dilemma of the classical impedance and admittance controllers. Stability of the impedance/admittance control is proven via Lyapunov-like analysis.

Another robot-interaction control scheme, human-in-the-loop control, is analyzed in Chapter 5. The inverse kinematics and the Jacobian matrix are avoided by using our Euler angle approach. This approach obtains a direct relation between the end-effector orientations and the joint angles. It is also applied in joint space and task space.

The second part of the book uses the reinforcement learning to the robot-interaction control. This part focuses more on optimal control and model-free control. Chapter 6 gives the design of position/force control using reinforcement learning. Chapter 7 extends it to large spaces and continuous spaces. The optimal desired impedance model, which minimizes the position error, is obtained by reinforcement learning. We use neural networks and the K-means clustering

Human-Robot Interaction Control Using Reinforcement Learning, First Edition. Wen Yu and Adolfo Perrusquía.
© 2022 The Institute of Electrical and Electronics Engineers, Inc. Published 2022 by John Wiley & Sons, Inc.

algorithm to approximate the large space. The convergence property is analyzed using the contraction property and Lyapunov-like analysis.

Finally we discuss three advanced nonlinear robot-interaction controllers using reinforcement learning. Chapter 8 is on the robust control under the worst-case uncertainty. Chapter 9 studies the redundant robots control with multi-agent reinforcement learning. Chapter 10 discusses robot $\mathcal{H}_2$ neural control.

The worst-case control is transformed into the $\mathcal{H}_2/\mathcal{H}_\infty$ problem and then into the form of reinforcement learning. In discrete time, we use the k-nearest neighbors and the double estimator technique to deal with the curse of dimensionality and the overestimation of the action values. By using reinforcement learning, the solutions of the inverse kinematics are obtained online for the redundant robots control. The multi-agent reinforcement learning uses two value functions to obtain the inverse kinematics. The $\mathcal{H}_2$ neural control uses recurrent neural networks to model the unknown robots, and then the optimal controller is designed in the sense of $\mathcal{H}_2$ and realized by reinforcement learning.

We make the following suggestions for future work:

- Learning from demonstrations. There is wide information on this control problem; however, there are many difficulties in theory and practice of the generalization of the training set.
- Deep reinforcement learning for robot-interaction control. This topic is relatively new and has become very popular in recent years.
- Human-robot interaction schemes. This book assumes the human does not have contact with the robot. Many human-robot interaction schemes need human impedance according to the environment, such as surgery with robots.

We believe that this book will help the new generations of scientists to explore this interesting research field and use the ideas discussed in the book to make new theoretical and practical contributions.

A

Robot Kinematics and Dynamics

A.1 Kinematics

The forward kinematic is the relation from the joint angles and the length of the links to the position and orientation of the end effector. The forward kinematic is defined as

$$x = f(q), \tag{A.1}$$

where $x \in \mathbb{R}^m$ is the position (X, Y, Z) and orientation (α, β, γ) of the robot end-effector, also known as Cartesian position, $x = (X, Y, Z, \alpha, \beta, \gamma)$, $q \in \mathbb{R}^n$ represents the joint angles position, and $f(\cdot)$ is the forward kinematics. The forward kinematics solution can be obtained using the well-known Denavit-Hartenberg notation [1].

The inverse kinematics is the relation from the position and orientation of the end effector, to the joint angles. The inverse kinematics is defined as

$$q = invf(x). \tag{A.2}$$

There is no standard or general method to solve the inverse kinematics $invf(\cdot)$ problem. Furthermore, there are multiple joint solutions that can achieve the same desired position and orientation (also known as pose) of the robot end effector.

Some well-known solution methods of the inverse kinematics problem are

1. Analytic solutions. Solutions are punctual and involve a low computational effort. They include: (a) Algebraic solution, b) Geometric solution.
2. Numerical solutions. They require a high computational effort due to algorithm iterations.

For redundant robots we can only have local solutions of the inverse kinematics, since we have to fix one or more joint angles in order to compute the others. So it is better to use the velocity kinematics approach.

Human-Robot Interaction Control Using Reinforcement Learning, First Edition. Wen Yu and Adolfo Perrusquía.
© 2022 The Institute of Electrical and Electronics Engineers, Inc. Published 2022 by John Wiley & Sons, Inc.

Jacobian

The Jacobian matrix helps us to avoid the inverse kinematics and provides useful properties for robot-interaction control schemes. The Jacobian is the derivative of the forward kinematics (A.1). It is the velocity mapping from the joint space (joint angles) to the Cartesian space (robot pose). The Jacobian is given by

$$\dot{x} = \frac{\partial f(q)}{\partial q} \dot{q} = J(q)\dot{q}, \tag{A.3}$$

where $\dot{x} \in \mathbb{R}^m$ are the Cartesian velocities, $\dot{q} \in \mathbb{R}^n$ are the joint velocities, and $J(q)$ is the Jacobian matrix.

There are two different Jacobians:

1. The analytic Jacobian, J_a, is the differential version of the forward kinematics (A.1).
2. The geometric Jacobian, J_g, is obtained from the Denavit-Hartenberg homogeneous matrices [1] and can be written as

$$J(q) = \begin{bmatrix} J_v(q) \\ J_\omega(q) \end{bmatrix}, \tag{A.4}$$

where $J_v(q) \in \mathbb{R}^{m' \times n}$ is the linear Jacobian, and $J_\omega(q) \in \mathbb{R}^{(m-m') \times n}$ is the angular Jacobian.

The relation between the analytic and the geometric Jacobian is

$$J_a = \begin{bmatrix} I & 0 \\ 0 & R(O)^{-1} \end{bmatrix} J_g, \tag{A.5}$$

where $R(O)$ is a rotation matrix of the orientation components: $O = [\alpha, \beta, \gamma]^\mathsf{T}$.

Virtual work principle

If a rigid body is under the action of external forces, the forces can be regarded as undergoing a virtual displacement, and these forces generate virtual work.

The virtual work is usually given by the dot product of the force (or torque) and the linear displacement (or angular displacement). According to the virtual work principle, these displacements are small:

$$F \cdot \delta x = \tau \cdot \delta q, \tag{A.6}$$

where $F \cdot \delta x$ is the virtual work of the end effector in task space, while $\tau \cdot \delta q$ is the virtual work of the joint displacements.

The above equation is also written as

$$F^\mathsf{T} \delta x = \tau^\mathsf{T} \delta q. \tag{A.7}$$

From the Jacobian definition (A.3), $\delta x = J(q)\delta q$, then

$$F^{\mathsf{T}} J(q)\delta q = \tau^{\mathsf{T}} \delta q. \tag{A.8}$$

So for all δq,

$$F^{\mathsf{T}} J(q) = \tau^{\mathsf{T}} \qquad or \qquad \tau = J^{\mathsf{T}}(q)F. \tag{A.9}$$

(A.9) shows the relation between the joints torque and the force in the end effector. This is one of the main relations for a robot-interaction control design.

Singularities

If $J(q)$ is not square or $\det \left[J(q) \right] = 0$, the Jacobian does not have full rank. In this case, it is called singularity or singular configuration. In the singular configuration, the robot dynamics loses a degree of freedom or loses the controllability.

When $\det \left[J(q) \right] \neq 0$, the inverse of the Jacobian is

$$J^{-1}(q) = \frac{1}{\det \left[J(q) \right]} \text{ adj} \left[J(q) \right].$$

To avoid the singularities from the complete Jacobian, we can use either the linear velocity Jacobian J_v, or the angular velocity Jacobian J_ω, or a combination of them. Therefore, we can avoid or add new singularity points depending on how we design the Jacobian matrix for the control problem.

A.2 Dynamics

The dynamics of a robot manipulator can be obtained through the Euler-Lagrange methodology. The kinetic energy of a robot manipulator is

$$\begin{aligned} \mathcal{K}(q, \dot{q}) &= \frac{1}{2}\dot{q}^{\mathsf{T}} M(q)\dot{q} \\ &= \frac{1}{2}\dot{q}^{\mathsf{T}} \underbrace{\left(\sum_{i=1}^{n} \left[m_i J_{v_i}^T(q)J_{v_i}(q) + J_{\omega_i}^{\mathsf{T}}(q)R_i(q)\mathcal{I}_i R_i^{\mathsf{T}}(q)J_{\omega_i}(q) \right] \right)}_{M(q)} \dot{q}, \end{aligned} \tag{A.10}$$

where m_i is the mass of link i, J_{v_i} and J_{ω_i} are the linear and angular Jacobian of link i respect to the other links, respectively, $\mathcal{I}_i$ is the inertia tensor of the link i, and $R_i(q)$ is the orientation matrix of each link with respect to the global inertial frame. $M(q) \in \mathbb{R}^{n \times n}$ is the inertia matrix, which is symmetric and positive definite for all $q \in \mathbb{R}^n$.

The potential energy is

$$\mathcal{U}(q) = \sum_{i=1}^{n} m_i g^{\mathsf{T}} r_{ci}, \tag{A.11}$$

where $g = \begin{bmatrix} g_x & g_y & g_z \end{bmatrix}^{\mathsf{T}}$ is the gravity direction with respect to the global inertial frame, and r_{c_i} are the coordinates of the center of mass of the link i.

The Lagrangian is defined as the difference between (A.10) and (A.11),

$$\mathcal{L}(q, \dot{q}) = \mathcal{K}(q, \dot{q}) - \mathcal{U}(q). \tag{A.12}$$

The complete form of the Euler-Lagrange equation (without dissipative terms) is

$$\frac{d}{dt}\left[\frac{\partial \mathcal{L}(q, \dot{q})}{\partial \dot{q}_i}\right] - \frac{\partial \mathcal{L}(q, \dot{q})}{\partial q_i} = \tau_i, \qquad i = 1, \cdots, n, \tag{A.13}$$

where τ_i are the external forces and torques applied to each joint.

The solution of the Euler-Lagrange equation (A.13) using the Lagrangian (A.12) is

$$\frac{d}{dt}(M(q)\dot{q}) - \frac{\partial \mathcal{K}(q, \dot{q})}{\partial q} + \frac{\partial \mathcal{U}(q)}{\partial q} = \tau,$$

$$M(q)\ddot{q} + \dot{M}(q)\dot{q} - \frac{\partial \mathcal{K}(q, \dot{q})}{\partial q} + \frac{\partial \mathcal{U}(q)}{\partial q} = \tau.$$

Let's define

$$C(q, \dot{q})\dot{q} \triangleq \dot{M}(q)\dot{q} - \frac{\partial \mathcal{K}(q, \dot{q})}{\partial q}, \qquad G(q) \triangleq \frac{\partial \mathcal{U}(q)}{\partial q}, \tag{A.14}$$

where $C(q, \dot{q}) \in \mathbb{R}^{n \times n}$ is the Coriolis and centripetal forces matrix, and $G(q) \in \mathbb{R}^n$ is the gravitational torques vector.

The terms of the Coriolis matrix can be reduced as

$$C(q, \dot{q})\dot{q} = \dot{M}(q)\dot{q} - \frac{\partial \mathcal{K}(q, \dot{q})}{\partial q} = \dot{M}(q)\dot{q} - \frac{1}{2}\dot{q}^{\mathsf{T}}\frac{\partial M(q)}{\partial q}\dot{q}$$

$$= \frac{1}{2}\dot{q}^{\mathsf{T}}\frac{\partial M(q)}{\partial q}\dot{q}, \qquad \text{where} \qquad \dot{M}(q) = \frac{\partial M(q)}{\partial q}\dot{q} = \dot{q}^{\mathsf{T}}\frac{\partial M(q)}{\partial q}.$$

The Coriolis matrix $C(q, \dot{q})$ is not unique. One way to compute the Coriolis matrix is by means of the Christoffel symbols method,

$$C_{kj} = \frac{1}{2}\sum_{i=1}^{n}\left\{\frac{\partial M_{kj}}{\partial q_i} + \frac{\partial M_{ki}}{\partial q_j} - \frac{\partial M_{ij}}{\partial q_k}\right\}\dot{q}_i. \tag{A.15}$$

Finally, the dynamic model of the robot manipulator with n DOFs is given by (A.17),

$$M(q)\ddot{q} + C(q, \dot{q})\dot{q} + G(q) = \tau. \tag{A.16}$$

If the interaction forces/torque is considered, the dynamic model (A.16) is

$$M(q)\ddot{q} + C(q, \dot{q})\dot{q} + G(q) = \tau - J^{\mathsf{T}}(q)f_e, \tag{A.17}$$

where $f_e = \begin{bmatrix} F_x & F_y & F_z & \tau_x & \tau_y & \tau_z \end{bmatrix}^{\mathsf{T}} \in \mathbb{R}^m$ are the forces and torques applied in different directions.

Dynamic model properties

The following special properties of the robot dynamics (A.16) [1, 104] are useful for control design and stability analysis.

There exist positive scalars $\beta_i(i = 0, 1, 2, 3)$ such that:

P1. The inertia matrix $M(q)$ is symmetric and positive definite and

$$0 < \beta_0 \le \lambda_{\min}\{M(q)\} \le \|M(q)\| \le \lambda_{\max}\{M(q)\} \le \beta_1 < \infty, \qquad (A.18)$$

where $\lambda_{\max}\{A\}$ and $\lambda_{\min}\{A\}$ are, respectively, the maximum and minimum eigenvalues of any matrix $A \in \mathbb{R}^{n \times n}$.
The norms $\|A\| = \sqrt{\lambda_{\max}(A^\top A)}$ and $\|b\|$ of vector $b \in \mathbb{R}^n$ stand for the induced Frobenius and vector Euclidean norms, respectively.

P2. For the Coriolis matrix $C(q, \dot{q})$,

$$\|C(q, \dot{q})\dot{q}\| \le \beta_2\|\dot{q}\|^2 \quad \text{or} \quad \|C(q, \dot{q})\| \le \beta_2\|\dot{q}\|, \qquad (A.19)$$

and $\dot{M}(q) - 2C(q, \dot{q})$ is skew symmetric, i.e.,

$$v^T \left[\dot{M}(q) - 2C(q, \dot{q})\right] v = 0 \qquad (A.20)$$

for any vector $v \in R^n$. Also,

$$\dot{M}(q) = C(q, \dot{q}) + C^\top(q, \dot{q}). \qquad (A.21)$$

P3. The gravitational torques vector $G(q)$ is Lipschitz

$$\left\|G(q_1) - G(q_2)\right\| \le k_g \left\|q_1 - q_2\right\| \qquad (A.22)$$

for some $k_g > 0$. Also it satisfies $\|G(q)\| \le \beta_3$.

P4. The Jacobian matrix is bounded. For a full-rank Jacobian, there exist positive scalars $\rho_i(i = 0, 1, 2)$ such that

$$
\begin{aligned}
\|J(q)\| &\le \rho_0 < \infty, \\
\|\dot{J}(q)\| &\le \rho_1 < \infty, \\
\|J^{-1}(q)\| &\le \rho_2 < \infty.
\end{aligned}
\qquad (A.23)
$$

Dynamic model in task space

Consider the Jacobian mapping (A.3) and its time derivative

$$
\begin{aligned}
\dot{x} &= J(q)\dot{q}, \\
\ddot{x} &= J(q)\ddot{q} + \dot{J}(q)\dot{q},
\end{aligned}
\qquad (A.24)
$$

where $\ddot{x} \in \mathbb{R}^m$ is the acceleration of the end effector, $\ddot{q} \in \mathbb{R}^n$ is the joints acceleration vector, and $\dot{J}(q) \in \mathbb{R}^{m \times n}$ is the time derivative of the Jacobian.

(A.24) gives the mapping between joint space and task (Cartesian) space. The robot interaction dynamics (A.17) in task space is

$$M_x \ddot{x} + C_x \dot{x} + G_x = f_\tau - f_e, \tag{A.25}$$

where

$$
\begin{aligned}
M_x &= J^{-T}(q)M(q)J^{-1}(q), \\
C_x &= J^{-T}(q)C(q, \dot{q})J^{-1}(q) - M_x(q)\dot{J}(q)J^{-1}(q), \\
G_x &= J^{-T}(q)G(q), \\
f_\tau &= J^{-T}(q)\tau.
\end{aligned}
\tag{A.26}
$$

All properties of the robot dynamics in joint space hold for the task space model.

A.3 Examples

Consider the following robots and systems used in this book.

Example A.1 *4-DOF exoskeleton robot*

Figure A.1 shows a 4-DOF exoskeleton robot. The robot has 3 DOFs that emulate the shoulder movement (flexion-extension, abduction-adduction, internal and external rotation) and 1 DOF that emulates the flexion-extension movement of the elbow. The DH parameters of the exoskeleton robot are given in Table A.1.

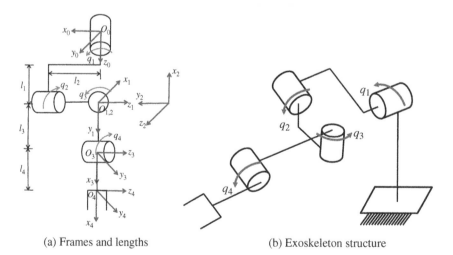

(a) Frames and lengths   (b) Exoskeleton structure

Figure A.1   4-DOF exoskeleton robot.

Table A.1 Denavit-Hartenberg parameters of the pan and tilt robot

Joint i	θ_i	d_i	a_i	α_i
1	q_1	l_1	0	$\frac{\pi}{2}$
2	q_2	0	0	$-\frac{\pi}{2}$
3	q_3	0	l_3	$\frac{\pi}{2}$
4	q_4	0	l_4	0

W $l_i, i = 1, 2, 3, 4$, are the lengths of each link, and $O_j, j = \overline{0,4}$ represents the origin of each inertial frame.

The homogeneous transformation matrix T of the exoskeleton is

$$T = \begin{bmatrix} c_1(c_2c_3c_4 - s_2s_4) - c_4s_1s_3 & s_1s_3s_4 - c_1(c_4s_2 + c_2c_3s_4) & c_3s_1 + c_1c_2s_3 & X \\ c_2c_3c_4s_1 + c_1c_4s_3 - s_1s_2s_4 & -c_4s_1s_2 - (c_2c_3s_1 + c_1s_3)s_4 & -c_1c_3 + c_2s_1s_3 & Y \\ c_3c_4s_2 + c_2s_4 & c_2c_4 - c_3s_2s_4 & s_2s_3 & Z \\ 0 & 0 & 0 & 1 \end{bmatrix},$$

where $c_i = \cos(q_i)$ and $s_i = \sin(q_i)$.

The Cartesian position of the robot is given by the forward kinematics which has the following form:

$$\begin{aligned} X &= c_1\left(c_2c_3(l_3 + l_4c_4) - l_4s_2s_4\right) - \left(l_3 + l_4c_4\right)s_1s_3, \\ Y &= c_2c_3\left(l_3 + l_4c_4\right)s_1 - l_4s_1s_2s_4 + c_1\left(l_3 + l_4c_4\right)s_3, \\ Z &= l_1 + c_3\left(l_3 + l_4c_4\right)s_2 + l_4c_2s_4. \end{aligned} \qquad \text{(A.27)}$$

The geometric Jacobian[1] is computed as follows:

$$J(q) = \begin{bmatrix} z_0 \times (O_4 - O_0) & z_1 \times (O_4 - O_1) & z_2 \times (O_4 - O_2) & z_3 \times (O_4 - O_3) \\ z_0 & z_1 & z_2 & z_3 \end{bmatrix}.$$

The exoskeleton geometric Jacobian is given by

$$J(q) = \begin{bmatrix} J_{11} & -c_1 J_{12} & J_{13} & J_{14} \\ J_{21} & -s_1 J_{12} & J_{23} & J_{24} \\ 0 & c_2c_3(l_3 + l_4c_4) - l_4s_2s_4 & -(l_3 + l_4c_4)s_2s_3 & l_4\left(c_2c_4 - c_3s_2s_4\right) \\ 0 & s_1 & -c_1s_2 & c_1c_2s_3 + c_3s_1 \\ 0 & -c_1 & -s_1s_2 & c_2s_1s_3 - c_1c_3 \\ 1 & 0 & c_2 & s_2s_3 \end{bmatrix}, \qquad \text{(A.28)}$$

with

$$J_{11} = -\left(l_3 + l_4 c_4\right)\left(c_2 s_1 + c_1 s_3\right) + l_4 s_1 s_2 s_4,$$
$$J_{12} = c_3(l_3 + l_4 c_4)s_2 + l_4 c_2 s_4,$$
$$J_{13} = -(l_3 + l_4 c_4)\left(c_3 s_1 + c_1 c_2 s_3\right),$$
$$J_{14} = l_4\left(s_1 s_3 s_4 - c_1 c_4 s_2 + c_2 c_3 s_4\right),$$
$$J_{21} = -(l_3 + l_4 c_4)s_1 s_3 + c_1\left(c_2 c_3(l_3 + l_4 c_4) - l_4 s_2 s_4\right),$$
$$J_{23} = (l_3 + l_4 c_4)\left(c_3 c_1 - s_1 c_2 s_3\right),$$
$$J_{24} = -l_4\left(s_1 s_2 c_4 + \left(c_2 c_3 s_1 + c_1 s_3\right)s_4\right).$$

The kinematic parameters of the exoskeleton robot according to the exoskeleton structure (see Figure A.1) are given in Table A.2.

Example A.2   2-DOF pan and tilt robot

Figure A.2 shows the 2-DOF pan and tilt structure and their respective inertial frames. The DH parameters of Table A.3 are obtained according to the inertial frames.

Table A.2   Kinematic parameters of the exoskeleton

Parameter	Description	Value
l_1	length of the shoulder socket	0.228 m
l_3	length of the arm	0.22 m
l_4	length of the forearm	0.22 m

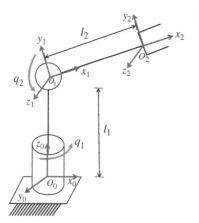

Figure A.2   2-DOF pan and tilt robot.

Table A.3 Denavit-Hartenberg parameters of the exoskeleton

Joint i	θ_i	d_i	a_i	α_i
1	q_1	l_1	0	$\frac{\pi}{2}$
2	q_2	0	l_2	0

The homogeneous transformation matrix T of the pan and tilt robot is

$$T = \begin{bmatrix} \cos(q_1)\cos(q_2) & -\cos(q_1)\sin(q_2) & \sin(q_1) & l_2\cos(q_1)\cos(q_2) \\ \sin(q_1)\cos(q_2) & -\sin(q_1)\sin(q_2) & -\cos(q_1) & l_2\sin(q_1)\cos(q_2) \\ \sin(q_2) & \cos(q_2) & 0 & l_1 + l_2\sin(q_2) \\ 0 & 0 & 0 & 1 \end{bmatrix}. \quad (A.29)$$

The forward kinematics is

$$\begin{aligned} X &= l_2\cos(q_1)\cos(q_2), \\ Y &= l_2\sin(q_1)\cos(q_2), \\ Z &= l_1 + l_2\sin(q_2). \end{aligned} \quad (A.30)$$

The geometric Jacobian is computed as follows:

$$\begin{aligned} J(q) &= \begin{bmatrix} z_0 \times (O_2 - O_0) & z_1 \times (O_2 - O_1) \\ z_0 & z_1 \end{bmatrix} \\ &= \begin{bmatrix} -l_2\sin(q_1)\cos(q_2) & -l_2\cos(q_1)\sin(q_2) \\ l_2\cos(q_1)\cos(q_2) & -l_2\sin(q_1)\sin(q_2) \\ 0 & l_2\cos(q_2) \\ 0 & \sin(q_1) \\ 0 & -\cos(q_1) \\ 1 & 0 \end{bmatrix}. \end{aligned} \quad (A.31)$$

For the dynamic model, we consider a diagonal inertial tensor, and its principal moments of inertia are assumed to be as if they were symmetric thin bars: $I_{xx_i} = I_{yy_i} = I_{zz_i} = \frac{m_i l_i^2}{12}$. The pan and tilt dynamics written as in (A.16) is

$$M(q) = \begin{bmatrix} M_{11} & 0 \\ 0 & M_{22} \end{bmatrix}, \quad C(q, \dot{q}) = \begin{bmatrix} C_1\dot{q}_2 & C_1\dot{q}_1 \\ -C_1\dot{q}_1 & C_2\dot{q}_2 \end{bmatrix}, \quad G(q) = \begin{bmatrix} 0 \\ G_2 \end{bmatrix}, \quad (A.32)$$

where

$$M_{11} = I_{yy_1} + \frac{1}{8}\left((4I_{yy_2} - 4I_{xx_2} + m_2 l_2^2) + \left(4I_{yy_2} - 4I_{xx_2} + m_2 l_2^2\right)\cos(2q_2)\right),$$

$$M_{22} = \frac{1}{4}\left(4I_{zz_2} + (l_1^2 + l_2^2)m_2 + 2l_1 l_2 m_2 \sin(2q_2)\right),$$

Table A.4 2-DOF pan and tilt robot kinematic and dynamic parameters

Parameter	Description	Value
m_1	mass of link 1	1 kg
m_2	mass of link 2	0.8 kg
l_1	length of link 1	0.0951 m
l_2	length of link 2	0.07 m
$I_{xx_1} = I_{yy_1} = I_{zz_1}$	moment of inertia of link 1	7.54×10^{-4} kgm^2
$I_{xx_2} = I_{yy_2} = I_{zz_2}$	moment of inertia of link 2	3.27×10^{-4} kgm^2

$$C_1 = -\frac{1}{8}\left(4I_{yy_2} - 4I_{xx_2} + m_2 l_1^2\right)\sin(2q_2),$$
$$C_2 = \frac{1}{4}l_1 l_2 m_2 \cos(q_2),$$
$$G_2 = \frac{1}{2}m_2 g l_2 \cos(q_2).$$

The kinematic and dynamic parameters of the pan and tilt robot are given in Table A.4, where $m_i, l_i, I_{xx_i}, I_{yy_i}, I_{zz_i}$, $i = 1, 2$, represent the mass, length, and moment of inertia at the X, Y, Z axes of the link i, respectively.

Example A.3 2-DOF planar robot

The dynamic model of a 2-DOF planar robot (see Figure A.3) with center of mass at the extreme of the link and written as (A.16) is

$$M(q) = \begin{bmatrix} M + m_2 l_2^2 + 2m_2 l_1 l_2 \cos(q_2) + J & m_2 l_2^2 + m_2 l_1 l_2 \cos(q_2) + J_2 \\ m_2 l_2^2 + m_2 l_1 l_2 \cos(q_2) + J_2 & m_2 l_2^2 + J_2 \end{bmatrix},$$

$$C(q, \dot{q}) = \begin{bmatrix} -m_2 l_1 l_2 \sin(q_2)\dot{q}_2 & -m_2 l_1 l_2 \sin(q_2)(\dot{q}_1 + \dot{q}_2) \\ m_2 l_1 l_2 \sin(q_2)\dot{q}_1 & 0 \end{bmatrix}, \quad \text{(A.33)}$$

$$G(q) = \begin{bmatrix} m_1 l_1 g \cos(q_1) + m_2 g\left(l_1 \cos(q_1) + l_2 \cos(q_1 + q_2)\right) \\ m_2 g\left(l_1 \cos(q_1) + l_2 \cos(q_1 + q_2)\right) \end{bmatrix}, \quad \tau = \begin{bmatrix} \tau_1 \\ \tau_2 \end{bmatrix},$$

where $M = (m_1 + m_2)l_1^2, q = [q_1, q_2]^T$, and m_i, l_i, J_i are the mass, length, and inertia of link i, with $i = 1, 2$ and $J = J_1 + J_2$.

Example A.4 Cart-pole system

The dynamics of the cart-pole system (see Figure A.4) written in the form of (A.16) is

$$M(q)\ddot{q} + C(q, \dot{q})\dot{q} + G(q) = Bu, \quad \text{(A.34)}$$

Figure A.3 2-DOF planar robot.

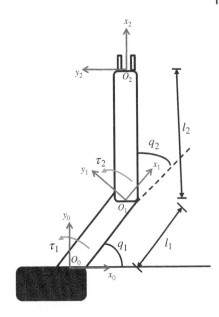

where

$$M(\mathbf{q}) = \begin{bmatrix} M + m & ml\cos(q) \\ ml\cos(q) & ml^2 \end{bmatrix}, \quad C(\mathbf{q}, \dot{\mathbf{q}}) = \begin{bmatrix} 0 & -ml\sin(q)\dot{q} \\ 0 & 0 \end{bmatrix},$$

$$G(\mathbf{q}) = \begin{bmatrix} 0 \\ -mgl\sin(q) \end{bmatrix}, \quad B = \begin{bmatrix} 1 \\ 0 \end{bmatrix},$$

(A.35)

where M is the cart mass, m is the pendulum mass, l is the pendulum length, g is the gravity acceleration, and F is the applied force at the cart. The generalized coordinates are represented as $\mathbf{q} = [x_c, q]^\mathsf{T}$, where x_c is the cart position and q is the angle formed between the pendulum and the vertical. This system is an under-actuated system whose control input is applied at the cart.

Figure A.4 Cart-pole system.

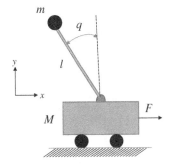

The system dynamics (A.35) can be rewritten in the form of (8.1) as follows

$$\dot{x} = \begin{bmatrix} 0 & 0 & 1 & 0 \\ 0 & 0 & 0 & 1 \\ 0 & 0 & 0 & 0 \\ 0 & 0 & 0 & 0 \end{bmatrix} x + \begin{bmatrix} 0 \\ 0 \\ \frac{cd_1 - bd_2}{ac - b^2} \\ \frac{ad_2 - bd_1}{ac - b^2} \end{bmatrix} + \begin{bmatrix} 0 \\ 0 \\ \frac{c}{ac - b^2} \\ \frac{b}{b^2 - ac} \end{bmatrix} u, \tag{A.36}$$

where $a = m + M$, $b = ml\cos(q)$, $c = ml^2$, $d_1 = ml\sin(q)\dot{q}^2$, and $d_2 = mgl\sin(q)$. The state vector is $x = (x_c, \dot{x}_c, q, \dot{q})^{\top}$, and $u = F$ is the control input.

References

1 M. W. Spong, S. Hutchinson, and M. Vidyasagar, *Robot Dynamics and Control.* John Wiley & Sons, Inc., 2nd ed., January 28, 2004.

2 O. A. Dominguez Ramirez, V. Parra Vega, M. G. Diaz Montiel, M. J. Pozas Cardenas, and R. A. Hernandez Gomez, *Cartesian Sliding PD Control of Robot Manipulators for Tracking in Finite Time:Theory and Experiments*, ch. 23, pp. 257–272. Vienna, Austria: DAAAM International, 2008.

3 R. Kelly and V. Santibáñez, *Control de Movimiento de Robots Manipuladores.* Ribera del Loira, España: Pearson Prentice Hall, 2003.

B

Reinforcement Learning for Control

B.1 Markov decision processes

In order to understand how dynamic programming (DP) works, we use the block diagram presented in Figure B.1. Here a controller interacts with the system using three signals: the system state, the action or control signal, and the scalar reward signal. The reward provides feedback on the controller performance. At each time step, the controller receives the states and applies the control action, which causes the system to change into a new state. This transition is given by the system dynamics. The behavior of the controller depends on its policy, which is a function from states into actions. This system-controller interaction with its states and actions constitutes a Markov decision process (MDP).

A deterministic MDP for control systems is defined by

$$MDP = (X, U, f, \rho),$$

where X is the state space of the system, U is the action space of the control, f is the system dynamics, and ρ denotes the reward function.

When the action $u_k \in U$ is applied in the state $x_k \in X$ at the discrete time k, the state is changed to x_{k+1}, according to the dynamics $f : X \times U \rightarrow X$:

$$x_{k+1} = f(x_k, u_k) \tag{B.1}$$

Simultaneously, the controller receives the scalar reward signal r_{k+1}, according to the reward function $\rho : X \times U \rightarrow \mathbb{R}$:

$$r_{k+1} = \rho(x_k, u_k) \tag{B.2}$$

The controller chooses actions according to its policy $h : X \times U$, using

$$u_k = h(x_k). \tag{B.3}$$

Human-Robot Interaction Control Using Reinforcement Learning, First Edition. Wen Yu and Adolfo Perrusquía.
© 2022 The Institute of Electrical and Electronics Engineers, Inc. Published 2022 by John Wiley & Sons, Inc.

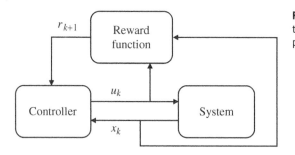

Figure B.1 Control system in the form of Markov decision process.

The main objective of dynamic programming (DP) is to find an optimal policy that minimizes the return, which is the cumulative value over the course of interaction. In the simplest case, the infinite-horizon return $R^h(x_k)$ is

$$\sum_{k=0}^{\infty} r_{k+1} = \sum_{k=0}^{\infty} \rho \left[x_k, h(x_k) \right] = R^h(x_k). \tag{B.4}$$

Another type of return uses the concept of discounting, where the controller tries to select actions or control actions so that the sum of discounted rewards it receives is minimized. The infinite-horizon discounted return is

$$\sum_{k=0}^{\infty} \gamma^k r_{k+1} = \sum_{k=0}^{\infty} \gamma^k \rho \left[x_k, h(x_k) \right] = R^h(x_k), \tag{B.5}$$

where $\gamma \in [0, 1)$ is the discount factor. Note that if $\gamma = 1$, (B.5) becomes (B.4).

The selection of γ often involves the trade-off between the quality of the solution and the convergence rate of the DP algorithm. If γ is too small, the solution may be unsatisfactory because it does not take into account rewards sufficiently.

When the controller-process interaction breaks naturally into subsequences, they are called episodes. Each episode ends in a terminal state, followed by a reset to a standard starting state. If the controller-process interaction does not break naturally, it defines a continuing task.

B.2 Value functions

There exist two types of value functions: (1) state value functions, i.e., V-functions and (2) state-action value functions, i.e., Q-functions. For any policy h, the Q-function, $Q^h : X \times U \rightarrow \mathbb{R}$ is

$$Q^h(x, u) = \sum_{k=0}^{\infty} \gamma^k \rho(x_k, u_k). \tag{B.6}$$

The above expression can be rewritten as

$$Q^h(x, u) = \rho(x, u) + \sum_{k=1}^{\infty} \gamma^k \rho(x_k, u_k) \tag{B.7}$$

$$= \rho(x, u) + \gamma R^h(f(x, u)),$$

where $\rho(x, u) = \rho(x_0, u_0)$. The optimal Q-function is defined as the best Q function that can be obtained by any control policy

$$Q^*(x, u) = \min_h Q^h(x, u). \tag{B.8}$$

The policy h^* is optimal if it selects at each state an action with the smallest optimal Q-value:

$$h^*(x) \in \operatorname*{argmin}_u Q^*(x, u). \tag{B.9}$$

In general, given a Q-function, the policy h is said to be greedy in Q if it satisfies

$$h(x) \in \operatorname*{argmin}_u Q(x, u). \tag{B.10}$$

According to Bellman equation, the value $Q^h(x, u)$ (taking action u in state x under the policy h) is equal to the immediate reward ρ and the discounted value by h in the next state:

$$Q^h(x, u) = \rho(x, u) + \gamma Q^h(x_{k+1}, u_{k+1})$$

$$= \rho(x, u) + \gamma Q^h \left[f(x, u), h(x_{k+1}) \right],$$

where $x_{k+1} = f(x_k, u_k)$, $u_k = h(x_k)$.
Bellman optimality equation for Q^* is

$$Q^*(x, u) = \rho(x, u) + \gamma \min_{u^*} Q^*(f(x, u), u^*). \tag{B.11}$$

The V-function $V^h : X \to \mathbb{R}$ of a policy h, is

$$V^h(x) = R^h(x) = Q^h(x, h(x)). \tag{B.12}$$

The optimal V-function is

$$V^*(x) = \min_h V^h(x) = \min_u Q^*(x, u). \tag{B.13}$$

The optimal control policy h^* can be computed from V^* as

$$h^*(x) \in \operatorname*{argmin}_u \left[\rho(x, u) + \gamma V^*(f(x, u)) \right]. \tag{B.14}$$

Since the V-function only describes the quality of the states, in order to infer the quality of transitions, V-functions use the following Bellman equations:

$$V^h(x) = \rho \left[x, h(x) \right] + \gamma V^h \left[f(x), h(x) \right], \tag{B.15}$$

$$V^*(x) = \min_u \left[\rho(x, u) + \gamma V^*(f(x, u)) \right]. \tag{B.16}$$

B.3 Iterations

In order to find the optimal value function and optimal policy, we need to apply the Bellman equations recursively. This recursion procedure is called iteration. The value iteration calculates the value function until the optimal value is obtained. The policy iteration evolves the policy until the optimal value function and the optimal control policy are achieved.

Let $\mathcal{L}$ be the set of all Q-functions and $\mathcal{H}$ be the value iteration, and compute the right-hand side of Bellman optimality equation

$$Q^*(x, u) = \rho(x, u) + \gamma \min_{u^*} Q^*(f(x, u), u^*). \tag{B.17}$$

For any Q-function, the value iteration is

$$\mathcal{H}[Q(x, u)] = \rho(x, u) + \gamma \min_{u^*} Q(f(x, u), u^*). \tag{B.18}$$

The Q-iteration algorithm starts from an arbitrary Q_0, and the Q -function is updated as

$$Q_{i+1} = \mathcal{H}[Q_i]. \tag{B.19}$$

Because $\gamma < 1$ in (B.18) and Q is applied to the infinity norm, it is true that

$$\|\mathcal{H}[Q_1] - \mathcal{H}[Q_2]\| \leq \gamma \|Q_1 - Q_2\| \tag{B.20}$$

for any Q_1 and Q_2.

To prove (B.20), we have

$$\mathcal{H}[Q_1] - \mathcal{H}[Q_2] = \rho(x, u) + \gamma \min_{u^*} Q_1(y, u^*) - \rho(x, u) - \gamma \min_{u^*} Q_2(y, u^*)$$

$$= \gamma \min_{u^*} \left(Q_1(y, u^*) - Q_2(y, u^*) \right)$$

$$\leq \gamma \|Q_1 - Q_2\|,$$

where y is any state.

It is common to use the max operator instead of the min operator:

$$\max_{u^*} Q(f(x, u), u^*) = -\min_{u^*} Q(f(x, u), u^*).$$

Bellman optimality equation states that Q^* is a fixed point on $\mathcal{H}$ and $\mathcal{H}$ has unique fixed point (equilibrium point),

$$Q^* = \mathcal{H}[Q^*]. \tag{B.21}$$

So the value iteration asymptotically converges to the optimal value function as

$$Q \to_{j \to \infty} Q^*.$$

Furthermore, Q-iteration converges to Q^* at a rate of γ,

$$\|Q_{j+1} - Q^*\| \leq \gamma \|Q_j - Q^*\|.$$

The policy iteration evaluates policies by constructing their value functions and uses these value functions to find new improved policies. The optimal policy can be computed from Q^*. Starting from the policy h_0, at every iteration j, the Q-function Q^{h_j} of the current policy h_j is determined. Policy evaluation is performed by solving Bellman equation.

When policy evaluation is complete, the new policy h_{j+1} is greedy found in Q^h

$$h_{j+1}(x) \in \min_u Q^{h_j}(x, u). \tag{B.22}$$

Policy iterations are also called policy improvement. The sequence of Q-functions produced by policy iteration asymptotically converges to Q^* as $j \to \infty$. Simultaneously, the optimal policy h^* is obtained.

B.4 TD learning

The temporal difference (TD) method can solve the Bellman equations online without knowing the full system dynamics. It is a model-free method. The control actions are calculated based on estimation of the value functions and the observing states.

The value iteration requires performing the recursion at each step, and the policy iteration requires the solution at each step of n linear equations. TD learning updates the value at each time step with the observations. The TD method adjusts values and actions online in real time along system trajectories.

Q-Learning

Q-learning, developed by Watkins and Werbos, is a TD method that is also called action-dependent heuristic dynamic programming. It works as follows:

1) Measure actual state x_k.
2) Apply an action u_k to the system.
3) Measure the next state x_{k+1}.
4) Obtain a scalar reward ρ.
5) Update the Q-function considering the current state, the action, the reward, and the next state $(x_k, u_k, \rho, x_{k+1})$.

The above process is the idea of reinforcement learning. It does not need the dynamic of the system. It is model-free MDP.

The value iteration of the Q-function is

$$\mathcal{H}[Q(x, u)] = \rho(x, u) + \gamma \min_{u^*} Q(f(x, u), u^*). \tag{B.23}$$

The Q-function is updated by the following Q-learning method,

$$Q_{k+1}(x_k, u_k) = Q_k(x_k, u_k)$$
$$+ \alpha_k \left\{ \rho(x_k, u_k) + \gamma \min_{u^*} \left[Q_k(x_{k+1}, u^*) \right] - Q_k(x_k, u_k) \right\}, \quad \text{(B.24)}$$

where $\alpha_k \in (0, 1]$, is the learning rate.

The term $\rho(x_k, u_k) + \gamma \min_{u^*} \left[Q_k(x_{k+1}, u^*) \right] - Q_k(x_k, u_k)$ is the temporal difference (TD) error. Since this learning is independent of the next policy h, it is an
off-policy algorithm. This value iteration algorithm can also arrive at the optimal
Q-function. To establish the convergence of Q-learning, the following lemma of
stochastic approximation is needed.

Lemma B.1 [1] *Consider the stochastic process* (ζ_k, Δ_k, F_k), $k \geq 0$, *where* ζ_k,
$\Delta_k, F_k : X \to \mathbb{R}$. *Let* P_k *be a sequence of increasing σ-fields, such that* ζ_0 *and* Δ_0 *are*
P_0 *-measurable, and* ζ_k, Δ_k, *and* F_k *are* P_k *-measurable; the process satisfies*

$$\Delta_{k+1}(z_k) = \Delta_k(z_k) \left[1 - \zeta_k(z_k) \right] + \zeta_k(z_k) F_k(z_k), \quad \text{(B.25)}$$

where $z_k \in X$, X *is finite. If the following conditions hold,*

1. $0 < \zeta_k(z_k) \leq 1, \sum_k \zeta_k(z_k) = \infty, \sum_k \zeta 2_k(z_k) < \infty, \forall z_1 \neq z_k : \zeta_k(z_1) = 0.$
2. $\|E\{F_k(x)|P_k\}\| \leq \kappa\|\Delta_k\| + c_k, \kappa \in (0, 1],$ *and* c_k *converges to zero a.s.*
3. $var\{F_k(z_k)|P_k\} \leq K(1 + \kappa\|\Delta_k\|)^2, K$ *is a positive constant,* $\|\cdot\|$ *denotes the*
   maximum norm.

then Δ_k *converges to zero with probability one.*

The following theorem gives the convergence of the Q-learning algorithm.

Theorem B.1 *For the finite MDP* (X, U, f, ρ), *the Q-learning algorithm (B.24) con-*
verges to the optimal value function Q^* *almost surely, if*

$$\sum_k \alpha_k = \infty, \quad \sum_k \alpha_k^2 < \infty. \quad \text{(B.26)}$$

Proof: The Q-learning algorithm (B.24) can be rewritten as,

$$Q_{k+1}(x_k, u_k) = (1 - \alpha_k)Q_k(x_k, u_k) + \alpha_k \left[\rho(x_k, u_k) + \gamma \min_{u^*} Q_k(x_{k+1}, u^*) \right], \quad \text{(B.27)}$$

where $r_{k+1} = \rho(x_k, u_k)$. Subtracting Q^* on both sides of (B.27) and defining the
value function error $\Delta_k(x_k, u_k) = Q_k(x_k, u_k) - Q^*$,

$$\Delta_{k+1}(x_k, u_k) = (1 - \alpha_k)\Delta_k(x_k, u_k) + \alpha_k \left(r_{k+1} + \gamma \min_{u^*} Q_k(x_{k+1}, u^*) - Q^* \right).$$

We define

$$F_k(x_k, u_k) = r_{k+1} + \gamma \min_{u^*} Q_k \left(x_{k+1}, u^* \right) - Q^*.$$

Use the value iteration mapping $\mathcal{H}$,

$$E\{F_k(x_k, u_k)|P_k\} = \mathcal{H} \left[Q_k(x_k, u_k) \right] - Q^*$$
$$= \mathcal{H} \left[Q_k(x_k, u_k) \right] - \mathcal{H}(Q^*).$$

Since $\mathcal{H}$ is a contraction,

$$\|E\{F_k(x_k, u_k)|P_k\}\| \leq \gamma \|Q_k(x_k, u_k) - Q^*(x_k, u_k)\| \leq \gamma \|\Delta_k(x_k, u_k)\|$$

Then

$$var\{F_k(x_k, u_k)|P_k\} = E\left[\left(r_{k+1} + \gamma \min_{u^*} Q_k(x_{k+1}, u^*) - \mathcal{H} \left[Q_k(x_k, u_k) \right] \right)^2 \right]$$
$$= var\left\{ r_{k+1} + \gamma \min_{u^*} Q_k(x_{k+1}, u^*)|P_k \right\}.$$

Because r_{k+1} is bounded, it clearly verifies

$$var\{F_k(x_k, u_k)|P_k\} \leq K \left(1 + \gamma \|\Delta_k(x_k, u_k)\| \right)^2,$$

where K is a constant. The second and the third conditions of of Lemma B.1 are satisfied. For Q-learning we usually select constant learning rate:

$$\alpha_k = \alpha, \quad 0 < \alpha \leq 1.$$

We choose

$$\alpha_k = \frac{\eta}{1 + \frac{1}{\beta} k}, \quad 0 < \eta \leq 1, \quad \beta \gg 1,$$

where η is a constant. Since β is very big, for finite time k, the learning rate, $\alpha_k \approx \eta$, is constant. Because

$$\sum_{k=1}^{\infty} \frac{\eta}{1 + \frac{1}{\beta} k} = \infty, \quad \sum_{k=1}^{\infty} \left(\frac{\eta}{1 + \frac{1}{\beta} k} \right)^2 = \eta^2 \beta^2 \psi(\beta, 1) - \eta^2 < \infty,$$

where $\psi(\beta, 1)$ is the digamma function, it is bounded. The first condition of Lemma B.1, or (B.26), is satisfied. Δ_k converges to zero w.p. 1, and hence, Q_k converges to Q^* with probability one. ∎

$\sum \alpha_k^2 < \infty$ in (B.26) requires that all state-action pairs are visited. This means the actions should be in every encountered state. The controller should select greedy actions to exploit all knowledge space. This is the exploration/exploitation trade-off in RL. A widely used method to balance exploration/exploitation is the

ε-greedy method: at each time step, instead of selecting greedily, a random action with fixed probability is used, $\varepsilon \in (0, 1)$,

$$u_k = \begin{cases} \text{random action in } U & \text{with uniform probability } \varepsilon_k, \\ u \in \text{argmin}_{u^*} Q_k(x_k, u^*) & 1 - \varepsilon_k. \end{cases} \qquad (B.28)$$

Sarsa

Sarsa is an alternative method for the value iteration. It joins every element in the data tuples $(x_k, u_k, r_{k+1}, x_{k+1}, u_{k+1})$: state (S), action (A), reward (R), (next) state (S), and (next) action (A). Start from any initial Q-function Q_0; on each step the data tuples are updated as follows:

$$Q_{k+1}(x_k, u_k) = Q_k(x_k, u) + \alpha_k \left[r_{k+1} + \gamma Q_k(x_{k+1}, u_{k+1}) - Q_k(x_k, u_k) \right]. \qquad (B.29)$$

In the Q-learning TD error includes the minimal Q-value in the next state. In Sarsa, the temporal difference error is $\left[r_{k+1} + \gamma Q_k(x_{k+1}, u_{k+1}) - Q_k(x_k, u_k) \right]$, and it includes the Q-value of the action in the next state. So Sarsa performs model-free policy evaluation.

Unlike the off-line policy iteration, Sarsa does not wait for the convergence of the Q-function; it uses unimproved policies to save time. Because of the greedy component, Sarsa implicitly performs policy improvement at every step. It is an online policy iteration. Sometimes it is called fully optimistic.

For the convergence proof, Sarsa needs similar conditions as Q-learning, which are established in the following theorem.

Theorem B.2   *Given a finite MDP (X, U, f, ρ), the values Q_k are computed by*

$$Q_{k+1}(x_k, u_k) = Q_k(x_k, u) + \alpha_k \left[r_{k+1} + \gamma Q_k(x_{k+1}, u_{k+1}) - Q_k(x_k, u_k) \right]. \qquad (B.30)$$

Then Q_k converges to the optimal Q^* and the policy $h(x_k)$ converges to the optimal policy $h^*(x_k)$ if the learned policy is greedy on the limit with infinite exploration (GLIE) and

$$\sum_k \alpha_k = \infty, \quad \sum_k \alpha_k^2 < \infty. \qquad (B.31)$$

Proof: The Sarsa update rule (B.29) can be rewritten as

$$\Delta_{k+1}(x_k, u_k) = (1 - \alpha_k)\Delta_k(x_k, u_k) + \alpha_k F_k(x_k, u_k),$$

where $\Delta_k(x_k, u_k) = Q_k(x_k, u_k) - Q^*$. F_k is given by

$$F_k(x_k, u_k) = r_{k+1} \pm \gamma \min_{u^*} Q_k(x_{k+1}, u^*) + \gamma Q_k(x_{k+1}, u_{k+1})$$

$$= r_{k+1} + \gamma \min_{u^*} Q_k(x_{k+1}, u^*) + C_k(x_k, u_k)$$

$$= F_k^Q(x_k, u_k) + C_k(x_k, u_k),$$

where

$$C_k(x_k, u_k) = \gamma \left(Q_k(x_{k+1}, u_{k+1}) - \min_{u^*} Q_k(x_{k+1}, u^*) \right),$$

$$F_k^Q(x_k, u_k) = r_{k+1} + \gamma \min_{u^*} Q_k(x_{k+1}, u^*) - Q^*.$$

Similarly with Q-learning,

$$E\{F_k^Q(x_k, u_k)|P_k\} \le \gamma \|\Delta_k(x_k, u_k)\|.$$

Let's define $Q_2(x_{k+1}, a) = \min_{u^*} Q_k(x_{k+1}, u^*)$, where a is the action that minimizes the function Q at the next state with $\Delta_k^a = Q_k(x, u) - Q_2(x, u)$. We consider two cases

1) If $E\left\{ Q_k(x_{k+1}, u_{k+1})|P_k \right\} \le E\{Q_2(x_{k+1}, a)|P_k\}$

$$Q_k(x_{k+1}, u_{k+1}) \le Q_k(x_{k+1}, a),$$

then

$$E\{C_k(x_k, u_k)|P_k\} = \gamma E\{Q_k(x_{k+1}, u_{k+1}) - Q_2(x_{k+1}, a)\}$$
$$\le \gamma E\{Q_k(x_{k+1}, a) - Q_2(x_{k+1}, a)\}.$$
$$\le \gamma \|\Delta_k^a\|.$$

2) If $E\{Q_2(x_{k+1}, a)|P_k\} \le E\{Q_k(x_{k+1}, u_{k+1})|P_k\}$,

$$Q_2(x_{k+1}, a) \le Q_2(x_{k+1}, u_{k+1}),$$

then

$$E\{C_k(x_k, u_k)|P_k\} = \gamma E\{Q_k(x_{k+1}, u_{k+1}) - Q_2(x_{k+1}, a)\}$$
$$\le \gamma E\{Q_k(x_{k+1}, u_{k+1}) - Q_2(x_{k+1}, u_{k+1})\}$$
$$\le \gamma \|\Delta_k^a\|.$$

Since the policy is GLIE, i.e., the non-greedy actions are chosen with vanishing probabilities. Hence, condition (B.26) is guaranteed. By means of Lemma B.1, Δ_k^a converges to zero,s and hence, Δ_k also converges to zero and Q_k converges to Q^*. ∎

Policy gradient

Q-learning uses the value of $Q(x, u)$ to take a certain action. It acts as a critic who evaluates the decision and the evaluation result using $Q_k(x_k, u_k)$. If the action space is continuous, the Q-learning algorithm needs to discretize the action space. The discretization will cause the action space to have a very high dimension, which makes it difficult for the Q-learning algorithm to find the optimal value, and the calculation speed is relatively slow.

The policy gradient makes up for this shortcoming directly by calculating what is the next action. That is, its output is the action or distribution of actions. It is like an actor that according to a certain state makes a certain action.

Let's denote by x_k the set of states, $R(x) = \sum_{k=0}^{\infty} r\left[x_k, u_k\right]$ the reward function, and $P(x, \theta)$ the probability of the trajectory x. The objective function is

$$J(\theta) = E\left[R(x), h(\theta)\right] = \sum_x P(x, \theta) R(x),$$

where $h(\theta)$ is the policy. The goal of reinforcement learning is to find the optimal parameter θ such that

$$\max J(\theta) = \max \sum_x P(x, \theta) R(x).$$

It uses the steepest descent method (gradient method) as

$$\theta_{k+1} = \theta_k + \eta \nabla J(\theta),$$

where

$$\nabla_\theta J(\theta) = \nabla_\theta \left(\sum_x P(x, \theta) R(x)\right) = \sum_x \nabla_\theta \left[P(x, \theta) R(x)\right]$$

$$= \sum_x P(x, \theta) \frac{\nabla_\theta\left[P(x, \theta)\right]}{P(x, \theta)} R(x) = \sum_x \nabla_\theta \left[\log P(x, \theta)\right] P(x, \theta) R(x)$$

$$= E\left[\nabla_\theta \left[\log P(x, \theta)\right] R(x)\right].$$

We can use empirical average to estimate $E\left[\nabla_\theta \left[\log P(x, \theta)\right] R(x)\right]$. If the current policy has m trajectories, the policy gradient with the empirical average of the m trajectories is approximated by

$$\nabla_\theta J(\theta) \approx \frac{1}{m} \sum_{k=1}^{m} \log P\left(x_k, \theta\right) R(x_k).$$

$\nabla_\theta \left[\log P(x, \theta)\right]$ is the gradient of the likelihood. Because

$$P(x, \theta) = \Pi_i P\left(x_{k+1}^i \mid x_k^i, u_k^i\right) h_\theta\left(x_k^i \mid u_k^i\right),$$

we have

$$\nabla_\theta \left[\log P(x, \theta)\right] = \sum \nabla_\theta \log P\left(x_{k+1}^i \mid x_k^i, u_k^i\right) + \sum \nabla_\theta \log h_\theta\left(x_k^i \mid u_k^i\right)$$

$$= \sum_i \nabla_\theta \log h_\theta\left(x_k^i \mid u_k^i\right).$$

Finally the the policy gradient is

$$\nabla_\theta J(\theta) = E\left[\nabla_\theta \left[\log h_\theta(x, \theta)\right] R(x)\right].$$

Actor-critic

The policy gradient is

$$\nabla_\theta J(\theta) = E\left[\nabla_\theta \left[\log h_\theta(x,\theta)\right] R(x)\right].$$

The index function of policy gradient is

$$J(\theta) = \sum_k \log h_\theta(x_k, u_k) R_k,$$

where R_k is the value function, h_θ is the action.

If R_k is estimated according to the Monte Carlo algorithm, it becomes

$$J(\theta) = \sum_k \log h_\theta(x_k, u_k) G_k.$$

G_k is obtained from multiple random variables. Another method is to introduce a baseline function,

$$J(\theta) = \sum_k \log h_\theta(x_k, u_k)\left[E(G_k) - R_k\right]$$
$$= \sum_k \log h_\theta(x_k, u_k)\left[Q_k(x_k, u_k) - R_k(x_k)\right],$$

where $E(G_k \mid x_k, u_k) = Q_k$. From Bellman equation,

$$Q_k(x_k, u_k) = E\left[r + \gamma R_k(x_{k+1})\right],$$
$$Q_k(x_k, u_k) = r_{k+1} + \gamma R_k(x_{k+1}).$$

So

$$J(\theta) = \sum_k \log h_\theta(x_k, u_k)\left[r + \gamma R_k(x_{k+1}) - R_k(x_k)\right].$$

Here $\left[r + \gamma R_k(x_{k+1}) - R_k(x_k)\right]$ can be regarded as the TD error, and it is estimated by the Q-learning algorithm.

The Actor-critic is a combination of Q-learning and policy gradient methods

1. The actor is the policy gradient algorithm to determine which action has the best effect.
2. The critic is Q-learning to evaluate an action taken in a certain state. It affects the actor's future choices.

The advantage of the actor-critic algorithm is the training time is shorter than the policy gradient.

Eligibility Trace

Eligibility trace is the basic mechanism of RL. It assists the learning process. It uses a short-term memory vector to assign the credits to the states visited.

At each step, the eligibility trace decays by $\lambda\gamma$, and for the one state visited it is denoted by $e_k(x)$,

$$
e_k(x) = \begin{cases} \lambda\gamma e_{k-1}(x) + 1 & \text{if } x = x_k, \\ \lambda\gamma e_{k-1}(x) & \text{if } x \neq x_k, \end{cases} \tag{B.32}
$$

where γ is the discount rate, and λ is the trace-decay parameter.

This eligibility trace is called accumulating trace because each time, the state is visited and faded gradually. A slight modification, such as the replacing trace, can significantly improve performance. The replacing trace for a discrete state x is defined as

$$
e_k(x) = \begin{cases} 1 & \text{if } x = x_k, \\ \lambda\gamma e_{k-1}(x) & \text{if } x \neq x_k. \end{cases} \tag{B.33}
$$

When the states are eligible for undergoing learning, the reinforcing events occur. Since we use one-step TD-error, the prediction of the state-value function is

$$
\delta_k = r_{k+1} + \gamma V_k(x_{k+1}) - V_k(x_k). \tag{B.34}
$$

This is the idea of the TD(λ) method. The TD error signals trigger proportionally to all recently visited states in the updating law as follows:

$$
W_{k+1}(x_k) = W_k(x_k) + \alpha_k \delta_k e_k(x). \tag{B.35}
$$

W_k is the parameter to be trained.

The accumulating and replacing traces are defined by the state-action pair,

$$
e_k(x, u) = \begin{cases} \lambda\gamma e_{k-1}(x, u) + 1 & \text{if } (x, u) = (x_k, u_k), \\ \lambda\gamma e_{k-1}(x, u) & \text{otherwise,} \end{cases} \tag{B.36}
$$

or

$$
e_k(x, u) = \begin{cases} 1 & \text{if } (x, u) = (x_k, u_k), \\ \lambda\gamma e_{k-1}(x, u) & \text{otherwise.} \end{cases} \tag{B.37}
$$

So the TD(λ) update for the Q-function becomes

$$
Q_{k+1}(x_k, u_k) = Q_k(x_k, u_k) + \alpha_k \delta_k e_k(x, u). \tag{B.38}
$$

Reference

1 A. Benveniste, M. Mativier, and P. Priouret, *Adaptive Algorithms and Stochastic Approximations*, vol. 22. Springer- Verlag: Berlin, 1990.

Index

2-DOF planar robot   158
4-DOF exoskeleton   83

2-DOF pan and tilt robot   83

a

accumulative traces   258
actor   154
actor learning rate   153
actor-critic methods   10
adaptive control   46
adaptive dynamic programming   193
adaptive law   47
admissible control policy   141
admittance control   5
algebraic Riccati equation   125
analytic Jacobian   79
approximators   6
asymptotic stability   46
augmented Jacobian method   177

b

Barbalat's lemma   198
basis function   122
Bellman equations   105
Bellman operator   122
Bellman optimal equation   105
Bellman's optimality principle   124

c

capacitive impedance   19
cart-pole system   154
cartesian nominal reference   49
classic linear controllers   154
continuous space   6
continuous-time actor-critic learning
       (CT-ACL)   153
continuous-time reinforcement learning
       (CT-RL)   124
contraction constant   122
contraction property   9
control policy   104
convergence   6
coriolis and centripetal forces matrix
       238
cost-index   25
critic   6
critic learning rate   153
curse of dimensionality   6

d

damping   21
damping ratio   40
dead zone   195
degrees of freedom (DOFs)   6
Denavit-Hartenberg   10
desired force   4

Human-Robot Interaction Control Using Reinforcement Learning, First Edition. Wen Yu and Adolfo Perrusquía.
© 2022 The Institute of Electrical and Electronics Engineers, Inc. Published 2022 by John Wiley & Sons, Inc.

desired reference 3
differential equation 17
discontinuous controllers 154
discount factor 102
discrete algebraic Riccati equation 103
discrete-time 6
distance metric 144
disturbances 3
double estimator 9
dual principle 19
dynamic programming 8

e
ε-greedy strategy 110
eligibility traces 8
end effector 21
environment 4
environment parameters 100
episode 106
Euclidean distance 144
Euler angles 8
Euler approximation 126
Euler-Lagrange 10
exoskeletons 6
expected value 144
exponential eligibility traces 129

f
feedback-feedforward control 196
fixed point 123
force control 4
force error 4
force/torque sensor 61
forward kinematics 9
fully cooperative stochastic game 181

g
Gaussian functions 128
geometric Jacobian 79
global asymptotic stability 51
gradient 25

gradient method 28
gravitational torques vector 238
gravity compensation 3
greedy policy 105

h
$\mathcal{H}_2$ control 10
$\mathcal{H}_2/\mathcal{H}_\infty$ control 139
$\mathcal{H}_\infty$ control 139
Hölder's inequality 128
Hamilton-Jacobi-Bellman 124
Hamilton-Jacobi-Isaacs 142
Hamiltonian 202
Hewer algorithm 125
human operator 8
human-in-the-loop control 8
human-robot interaction 86
Hurwitz matrix 194
hybrid control 4
hybrid reinforcement learning 9
hyper-parameters 154

i
impedance control 4
impedance model 5
inertia matrix 237
inertial impedance 19
infinite-horizon discounted return 248
infinite-horizon return 248
inner loop position control 74
input space 6
input-state stability 46
integral reinforcement learning 125
integral squared error (ISE) 166
inverse kinematics 9

j
Jacobian matrix 9
joint displacements 9
joint space 3

k

k nearest neighbors 9
K-means clustering 9
Kelvin-Voigt model 24
Kronecker product 196

l

Lagrangian 238
Laplace transform 17
large state and discrete actions (LS-DA) 146
large state and large action (LS-LA) 146
learning rate 28, 106
learning time 7
least-squares method 25
Leibniz rule 124
linear combination 76
linear quadratic regulator 3
linear system theory 3
linear time-variant system (LVT) 205
linearization 76
Lipschitz condition 208
localizable functions 140
long short-term memory 193
Lyapunov equation 196
Lyapunov recursions 125
Lyapunov stability theory 8

m

machine learning 6
Markov decision process 105
mechanical impedance 19
model compensation 3
model pre-compensation 3
modeling error 6
Monte Carlo 6
Moore-Penrose pseudoinverse 176
Multi-agent reinforcement learning 9

n

natural frequency 40
near-optimal performance 136

neural control 3
nominal reference 36
normalized gradient descent algorithm 202
normalized radial basis functions 120
Norton equivalent circuit 19
null space 175

o

off-policy algorithm 106
online policy algorithm 106
one-step update 154
optimal control policy under constraints 104
optimal performance 4
optimal value function 104
optimization problem under constraints 9
orientation components 78
outer loop force control 74
overestimation problem 9

p

parallel recurrent neural network 208
parameter convergence 28
parameter identification 24
parameters 7
parametric error 28
policy gradient 256
policy search 6
policy-iteration 121
position control 4
position error 3
position tracking 101
position/force control 9
precision and robustness problem 8
predictor 144
probability distribution 145
projection function 122
Proportional-Derivative 3
Proportional-Integral-Derivative 3
pseudoinverse method 173

q
$Q(\lambda)$ 107
Q-function   105
Q-learning   6

r
radial basis functions   9
range space   175
recurrent neural networks   193
recursive least squares   26
redundant robot   9
regression matrix   24
reinforcement learning   3
replacing traces   258
residual error   202
resistive impedance   19
reward function   6
robot dynamics   3
robust reward   143
robustness   5

s
Sarsa   6
Sarsa(λ)   107
semi-global stability   48
serial learning   147
serial-parallel recurrent neural network
    194
set-point control   174
signum function   51
single hidden layer feedforward neural
    network (SLFNN)   177
singular value decomposition (SVD)
    176
singularities   9

singularity avoidance   173
sliding mode control   3
sliding surface   51
spring-damper system   24
steady-state error   50
stiffness   21
stochastic game   181
Stone-Weierstrass theorem   194
super-twisting sliding mode control   51

t
task space   8
Taylor expansion   195
TD-error   106
temporal difference methods   8
Thèvenin equivalent circuit   19
total cumulative reward   139
training steps   110

u
unbiased estimators   145
uniform ultimate bounded (UUB)   48
utility function   104

v
value function   103
velocity kinematics   9
virtual work principle   52

w
weight matrices   102
worst-case uncertainty   9

z
zero-sum differential game   140

IEEE PRESS SERIES ON SYSTEMS SCIENCE AND ENGINEERING

Editor:
MengChu Zhou
New Jersey Institute of Technology and Tongji University

Co-Editors:
Han-Xiong Li
City University of Hong-Kong

Margot Weijnen
Delft University of Technology

The focus of this series is to introduce the advances in theory and applications of systems science and engineering to industrial practitioners, researchers, and students. This series seeks to foster system-of-systems multidisciplinary theory and tools to satisfy the needs of the industrial and academic areas to model, analyze, design, optimize and operate increasingly complex man-made systems ranging from control systems, computer systems, discrete event systems, information systems, networked systems, production systems, robotic systems, service systems, and transportation systems to Internet, sensor networks, smart grid, social network, sustainable infrastructure, and systems biology.

1. *Reinforcement and Systemic Machine Learning for Decision Making*
   Parag Kulkarni
2. *Remote Sensing and Actuation Using Unmanned Vehicles*
   Haiyang Chao and YangQuan Chen
3. *Hybrid Control and Motion Planning of Dynamical Legged Locomotion*
   Nasser Sadati, Guy A. Dumont, Kaveh Akbari Hamed, and William A. Gruver
4. *Modern Machine Learning: Techniques and Their Applications in Cartoon Animation Research*
   Jun Yu and Dachen Tao
5. *Design of Business and Scientific Workflows: A Web Service-Oriented Approach*
   Wei Tan and MengChu Zhou
6. *Operator-based Nonlinear Control Systems: Design and Applications*
   Mingcong Deng
7. *System Design and Control Integration for Advanced Manufacturing*
   Han-Xiong Li and XinJiang Lu

8. *Sustainable Solid Waste Management: A Systems Engineering Approach*
   Ni-Bin Chang and Ana Pires
9. *Contemporary Issues in Systems Science and Engineering*
   MengChu Zhou, Han-Xiong Li, and Margot Weijnen
10. *Decentralized Coverage Control Problems For Mobile Robotic Sensor and Actuator Networks*
    Andrey V. Savkin, Teddy M. Cheng, Zhiyu Li, Faizan Javed, Alexey S. Matveev, and Hung Nguyen
11. *Advances in Battery Manufacturing, Service, and Management Systems*
    Jingshan Li, Shiyu Zhou, and Yehui Han
12. *Automated Transit: Planning, Operation, and Applications*
    Rongfang Liu
13. *Robust Adaptive Dynamic Programming*
    Yu Jiang and Zhong-Ping Jiang
14. *Fusion of Hard and Soft Control Strategies for the Robotic Hand*
    Cheng-Hung Chen, Desineni Subbaram Naidu
15. *Energy Conservation in Residential, Commercial, and Industrial Facilities*
    Hossam A. Gabbar
16. *Modeling and Control of Uncertain Nonlinear Systems with Fuzzy Equations and Z-Number*
    Wen Yu and Raheleh Jafari
17. *Path Planning of Cooperative Mobile Robots Using Discrete Event Models*
    Cristian Mahulea, Marius Kloetzer, and Ramón González
18. *Supervisory Control and Scheduling of Resource Allocation Systems: Reachability Graph Perspective*
    Bo Huang and MengChu Zhou
19. *Human-Robot Interaction Control Using Reinforcement Learning*
    Wen Yu and Adolfo Perrusquía

Printed and bound by CPI Group (UK) Ltd, Croydon, CR0 4YY